Fluid Mechanics and Thermodynamics of Turbomachinery

Fluid Mechanics and Thermodynamics of Turbomachinery

Editor

Prabhu Nirajan

scitus
academics

Fluid Mechanics and Thermodynamics of Turbomachinery
Edited by **Prabhu Nirajan**

Printed in 2017

ISBN: 978-1-68117-375-7

Library of Congress Control Number: 2015941563

Notice

Reasonable efforts have been made to publish reliable data and views articulated in the chapters are those of the individual contributors, and not necessarily those of the editors or publishers. Editors or publishers are not responsible for the accuracy of the information in the published chapters or consequences of their use. The publisher believes no responsibility for any damage or grievance to the persons or property arising out of the use of any materials, instructions, methods or thoughts in the book. The editors and the publisher have attempted to trace the copyright holders of all material reproduced in this publication and apologize to copyright holders if permission has not been obtained. If any copyright holder has not been acknowledged, please write to us so we may rectify.

Contents

Preface

Turbomachinery is a challenging and diverse field, with applications for professionals and students in many subsets of the mechanical engineering discipline, including fluid mechanics, combustion and heat transfer, dynamics and vibrations, as well as structural mechanics and materials engineering. Originally published more than 40 years ago, Fluid Mechanics and Thermodynamics of Turbomachinery is the leading turbomachinery textbook. Used as a core text in senior undergraduate and graduate level courses this book will also appeal to professional engineers in the aerospace, global power, oil & gas and other industries who are involved in the design and operation of turbomachines. Turbomachinery is a challenging and diverse field, with applications for professionals and students in many subsets of the mechanical engineering discipline, including fluid mechanics, combustion and heat transfer, dynamics and vibrations, as well as structural mechanics and materials engineering.

Editor

Chapter 1

The Analysis of the Noise Generation in Gas Turbine Stage

Sławomir Dykas, and Dawid Machalica

Institute of Power Engineering and Turbomachinery, Silesian University of Technology, Gliwice, Poland

ABSTRACT

The aim of this paper is to assess the impact of the mutual positioning of the turbine stage stator and rotor blades on noise generation. The Ansys CFX commercial software package and the Scale-Adaptive Simulation (SAS) hybrid turbulence model are used for numerical analyses. The paper is focused on an analysis that the pressure wave generation resulting from unsteady flow phenomena. In order to present the problem, the Fast Fourier Transformation (FFT) analysis of pressure fluctuation is carried out at selected points of the turbine stage computational domain. A comparison of values of individual components for subsequent control points allows an approximate determination of the place of generation of pressure waves, the direction of their propagation and the damping rate. Moreover, the numerical analyses make it possible to evaluate the justification for the use of the SAS model, which is rather demanding in terms of equipment, in simulations of unsteady flow fields where generation and propagation of noise waves occur.

INTRODUCTION

The rapid development of numerical methods and techniques in the last two decades has made it possible to carry out more and more complex analyses using the Computational Fluid Dynamics (CFD) tools. Owing to the use of computer software in flow analyses, it is possible to avoid high costs related to the construction of test stations and reduce the time needed to obtain the optimum structure. It is also significant that simulation results contain full information on the flow medium properties which under experimental conditions often cannot be measured (e.g. density or entropy). These advantages contribute to the fact that numerical analyses are gaining popularity in industries such as power engineering or aviation. This, however, does not mean that experimental testing should be discontinued as it is necessary to validate numerical models.

A correct assessment of the turbine stage operation should be made through an analysis of unsteady phenomena which provides lots of information concerning instantaneous distributions of pressures and other flow parameters of the medium. The knowledge of these parameters is useful while analysing vibration or acoustic phenomena. In extreme situations they may lead to destruction of the machine components due to high cycle wear. The knowledge of pressure fluctuation at individual points of the blade channel makes it also possible to perform analyses of generation and propagation of pressure waves, including acoustic waves (noise or/and vibrations).

Nowadays, this is of special importance considering the ever-stricter restrictions on the noise level of, among others, aircraft engines. Because of the binding standards, manufacturers need to optimise the structure in terms of noise generation as early as at the designing stage. Only then will customers become interested in the supplier's products. Unsteady flow field analyses also allow, for example, an assessment of operation during a change in load. This gives an idea of the phenomena occurring inside the machine which quite often cannot be analysed experimentally.

The unsteady phenomena taking place in channels of turbomachines in many cases result in energy dissipation. Any phenomenon leading to a drop in efficiency is a loss [1] . Analysis of the losses in turbine stages by means various loss coefficients was performed by many authors [1] - [4]. However, the process of the acoustic wave generation in the flow field may also be perceived as a measure of energy dissipation.

This paper presents the methodology and the results of CFD analyses comprising the flow in the turbine stage of an engine. The topic was covered fairly often [5] [6] . However, the authors focused on selected issue only. For this purpose, the Ansys CFX commercial software is used. Due to anticipated complexity of the flow field, a decision was made to employ the Scale-Adaptive Simulation (SAS) hybrid turbulence model. This model can adapt itself dynamically to the already solved vortex street [7] [8]. Owing to that, it works perfectly in separation areas for example. In the paper, the focus is on the FFT analysis of pressure fluctuation at selected points of the computational domain. Comparing the values of varying pressure, it is possible to roughly determine the places of generation of acoustic waves, the direction in which they propagate and their damping rate.

ANALYZED TURBINE STAGE

The object selected for the numerical analyses whose results are presented in this paper is the stage of a gas turbine intended for the aircraft industry [9] [10] . This turbine features a single stage only. A reduction in the number of stages involves a rise in the thermal and pressure drop per a turbine stage. This has an impact on the strength and flow properties. Moreover, the turbine under consideration was to operate ultimately at the temperature of the flue gases supplied to the stator at the level of 2200 K. Such high temperature, unusual even in modern structures, required on part of the designers a development of cooling systems that would prevent blades from melting. In this paper, however, the issues related to blade cooling are ignored to reduce the size of the numerical mesh and to shorten the time of computations.

The analysed turbine is composed of 36 38.1 mm high stator blades located at an average radius of 469.9 mm and 64 rotor blades with the same height and radius of location. The computations are performed for one stator channel and two rotor channels. Due to the lack of data concerning the geometry of the tip seal of the rotor blades, an example typical geometry is selected corresponding to two fins and a clearance of 0.3 mm. The geometry of the stage under analysis is presented in Figure 1.

PHYSICAL AND NUMERICAL

The numerical analyses in this paper are carried out using the Ansys CFX commercial package using the uRANS method and the Scale-Adaptive Simulation (SAS) turbulence model. This model belongs to the group of hybrid models. It can adapt itself dynamically to the already solved vortex structures yielding good results in areas with intense swirling. It thus constitutes a compromise between the classic viscosity model and the DES or LES methods, which are rather costly in terms of computations and which are also available in the Ansys CFX package [4] [6] .

The unsteady computations comprised almost 2000 time steps distanced from each other by 10^{-5}s. This made it possible to carry out FFT analyses using 1024 samples, excluding the beginning stage steps which comprise the time for the solution to stabilize.

Additionally, a short analysis was performed with a time step of 3 × 10^{-6} s, which allows the assessment of the changes in the values of loss coefficients depending on the position of the stator and rotor blades.

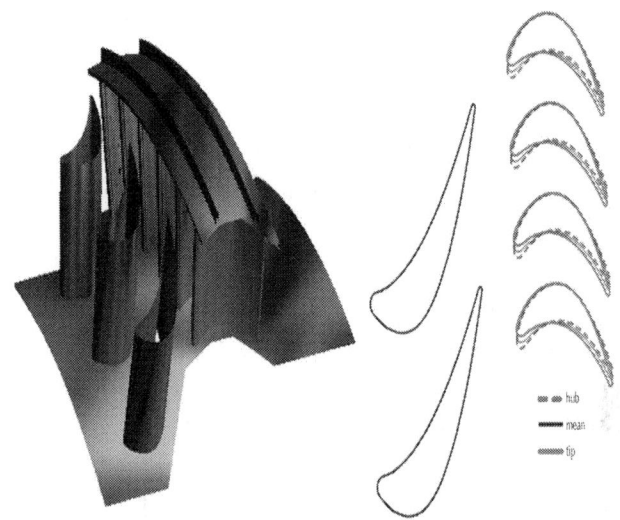

Figure 1: Geometry of the analysed turbine stage.

The desire to model the turbine stage together with the rotor blade tip seal requires a division of the computational area into three parts: the stator channel, the rotor channel and the seal area. For each of these parts a separate numerical mesh was generated. This accelerated the process of mesh generation and improved the mesh quality. The meshes for the flow channel around the stator and the rotor are of the mixed type, where the maximum value of the non-dimensional parameter y+ describing the size of the mesh near-wall element was approximately 1.20.

The seal mesh was generated in the ICEM CFD program as fully structured. In order to improve the accuracy of transition modeling at the stator/rotor interface, two rotor channels were assumed for the calculations, which in total gave more than 4.7 M control volumes of the mesh for the entire stage with the seal (Figure 2).

The numerical analysis was carried out for the turbine design parameters described in a NASA report [10]. The characteristic feature of the turbine is that it has just one stage, which distinguishes it from other turbines used in turbofan engines. In order to achieve a required load the proposed value of the total temperature was 2200 K, what is not reachable till nowadays. For reasons of simplicity, the medium fed to the stage is treated as perfect gas. The remaining boundary conditions used in the analysis are presented in Table 1.

NUMERICAL RESULTS—FFT ANALYSIS

An FFT analysis was carried out at selected points of the computational area in order to estimate the places where pressure waves are generated and the direction in which they propagate. The location of the points discussed in this paper is presented in Figure 3.

The first element of the name of the control point Ro/U (Rotor/ Seal) identifies the computational domain where the point is located. The element in/out/o identifies the axial location of the point with respect to the blade. The element b/ch indicates whether the point is located before the leading edge/in the vortex after the blade (b) or at the blade channel inlet/outlet (ch). This particular arrangement is to make it possible to estimate the distribution of pressure fluctuation in the axial and circumferential directions. Moreover, the points are located at three heights of 20%, 50% and 80% of the relative height of the blade channel.

Ahead of the Rotor

For points located before the rotor blade (Figure 4(a)), in the vicinity of the impact point, components p' feature quite similar amplitudes. The highest value of p' for the 13.2 kHz component occurred in the bottom part of the channel; the lowest—in its middle part, where the effect of the walls restricting the flow is the smallest. It is also worth noting that despite the turbine very high rotational speed, the generated components with a high amplitude are still included in the audible frequency range.

The situation is identical at the rotor channel inlet (Figure 4(b)), where the distribution of p′ amplitudes is the same as before the impact point. However, at the rotor channel inlet the amplitude values are higher by about 20 kPa. The increase in amplitudes inside the channel is probably caused by the interference of waves arising at impact points of the blades located nearby.

In the case of points located before the rotor blade, apart from the 13.2 kHz component, there is also a component with f ≈ 8.6 kHz, and more components corresponding to it with frequencies increasing by approx- imately the Vane Passing Frequency (VPF), which in this case is ~13 kHz.

Figure 2: Computational meshes of the stator and rotor channels and of the seal.

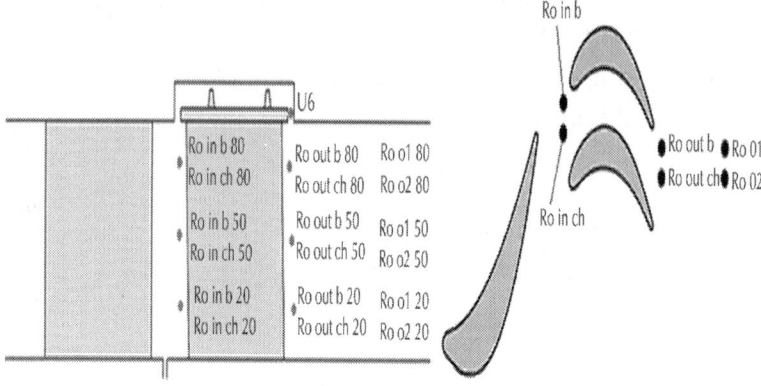

Figure 3: Arrangement of control points in the computational area.

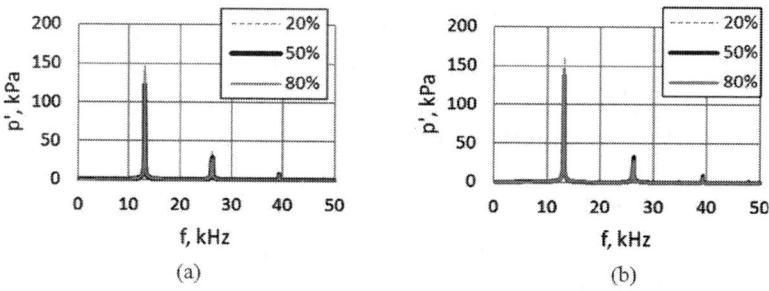

Figure 4: FFT analysis results for control points located before: (a) The rotor blade (points "Ro in b"); (b) The blade channel inlet (points "Ro in ch").

Table 1: Data assumed for the real analysis

	Unsteady analysis
Medium	Air as perfect gas
Turbulence model	SAS
Rotor rotational speed	21,772 RPM
Stator/rotor interface type	Transient rotor-stator

Inlet	Total pressure	3.861 MPa
	Static temperature	2200 K
	Turbulence intensity	5%
Outlet	Static pressure	600 kPa
	Radial equilibrium	Yes

It becomes especially visible if p′ on the y axis is substituted with SPL (Figure 5). Here, each subsequent component features a lower amplitude reaching as much as 160 dB for f = 47.9 kHz.

Seal Outlet

The medium flow through the seal is extremely complex. The consequences of the very turbulent nature of the flow in this area can be observed especially at the outlet, where point U6 is located. The FFT analysis carried out for control point U6 (Figure 6) confirms this observation. The SAS model made it possible to capture a large number of new components than others two-equation eddy-viscosity turbulence models (for example the Shear Stress Transport (SST) model), including very significant ones with frequencies of 0.8 kHz, 3.9 kHz and 7.6 kHz. The components appearing in the range of high frequencies may have their source both in the occurrence in the flow of vortex structures featuring small time scales and in the generation of acoustic waves arising due to the thermoacoustic effect. They would be induced by the temperature difference between the medium flowing out of the necking in the seal (over the fin) and the medium with a different temperature included in the seal cavity.

Points Located before the Rotor Blade Inflow Edge and after the Blade, in the Aerodynamic Wake

Comparing the results obtained for points Ro in b 80 and Ro out b 80 (Figure 7), a distinct reduction in the amplitude fluctuation by the VPF can be noticed for the point located after the blade. The reduction is almost 80-fold. It can also be seen that after the blade, at the height of 80%, a component can be detected with a frequency of 3.9 kHz, which is characteristic of the medium leaving the seal. It reaches ~400 Pa and is thus about eight times smaller than at the seal outlet.

Points Located after the Rotor Row

After the rotor blade row, both at points located in the aerodynamic wake (Figure 8) and at the rotor channel outlet (Figure 9), amplitude p' for the 13.2 kHz component diminishes with the blade channel height. This change is about fivefold for points located at the rotor channel outlet and 40-fold for points in the blade wake.

In the range of low frequencies, the components which are discernible first of all are those coming from the seal (3.9 kHz), which are present only at control points located at 80% of the channel height. The amplitude of a component featuring this frequency is three times higher at a point after the blade than at the blade channel outlet. At the point located at the rotor channel outlet there is an additional component with $f \approx 800$ Hz, which was previously also noticed at point U6.

At points located after the rotor row, more additional components can be observed in the range of frequencies higher than 30 kHz (e.g. 43.5 and 44 kHz). These components feature amplitudes of up to 150 Pa (the compo- nent being a VPF multiple ignored).

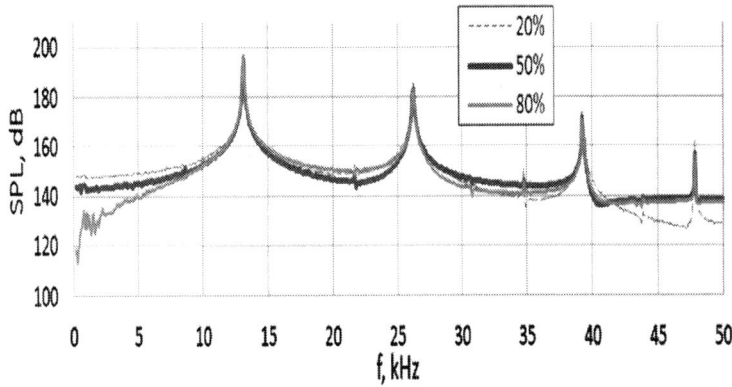

Figure 5: FFT analysis results for control points located before the rotor blade (points "Ro in b").

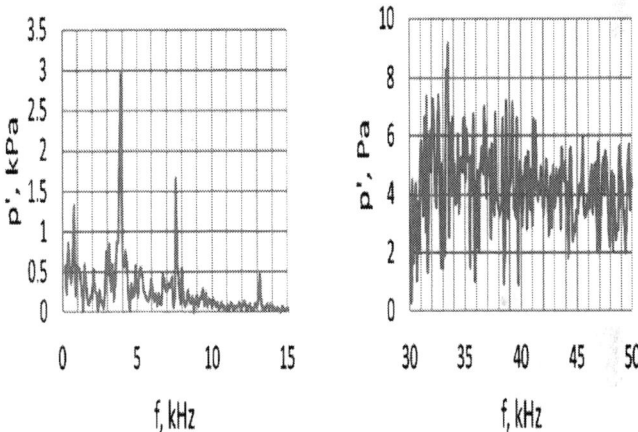

Figure 6: Results of the FFT analysis carried out for control point U6.

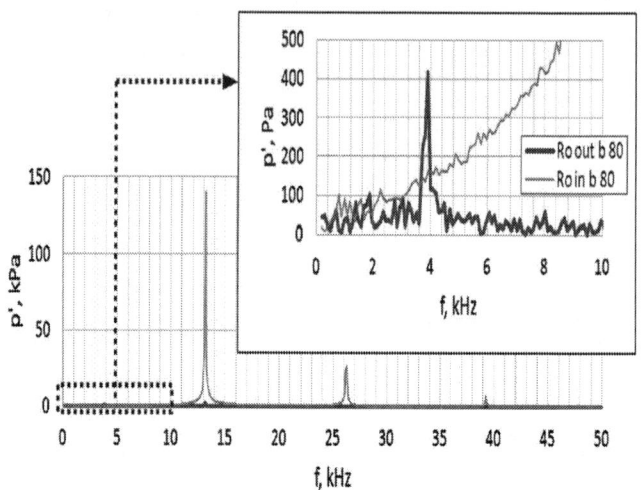

Figure 7: FFT analysis results for control points located before and after the rotor blade.

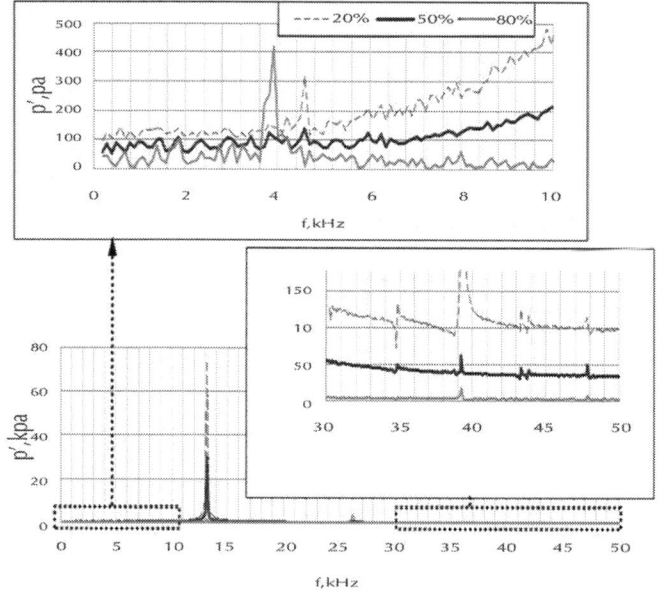

Figure 8: FFT analysis results for control points located after the rotor blade (points "Ro out b").

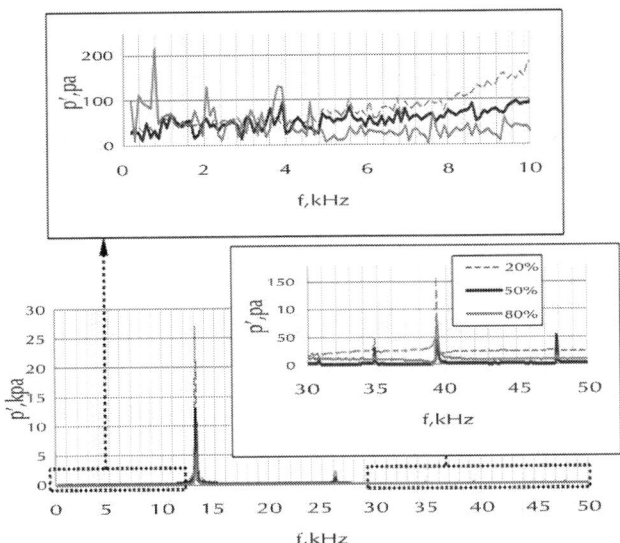

Figure 9: FFT analysis results for control points located at the rotor channel outlet (points "Ro out ch").

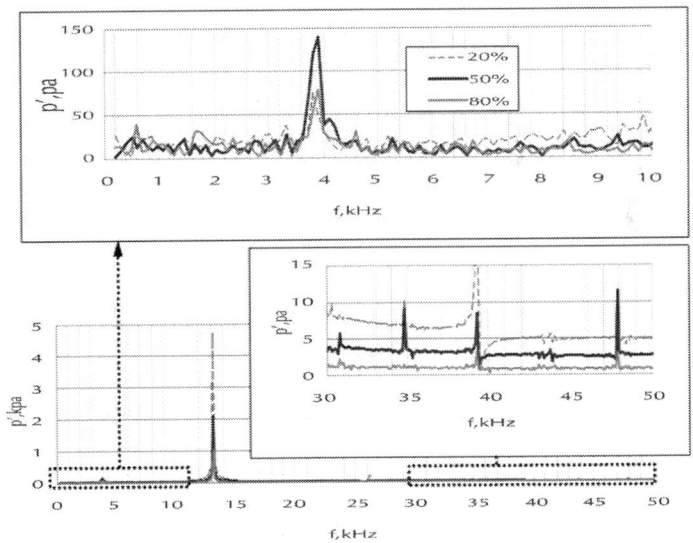

Figure 10: FFT analysis results for control points Ro o1 80, Ro o1 50, Ro o1 30.

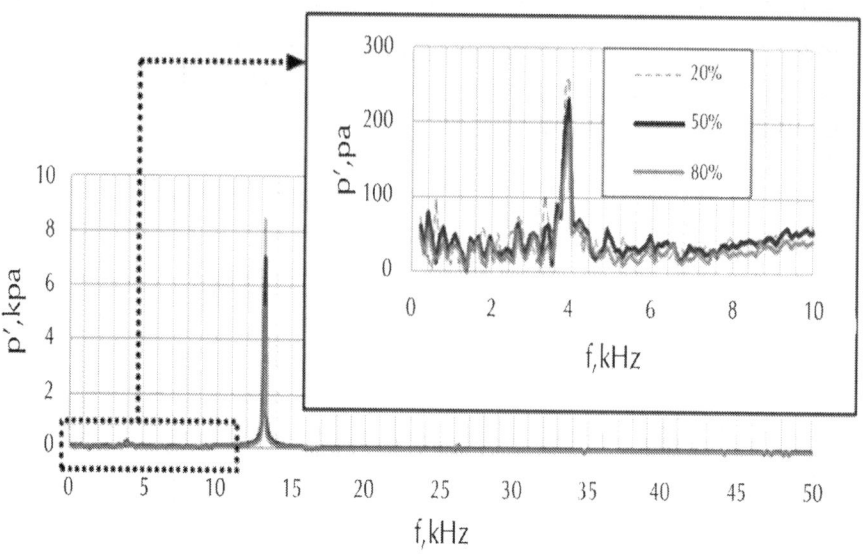

Figure 11: FFT analysis results for control points Ro o2 80, Ro o2 50, Ro o2 30.

They may prove the impact of the thermoacoustic effect on generation of acoustic waves in the area of the trailing edge because a very high temperature gradient occurs there between the medium flowing out of the blade channel and the medium in the aerodynamic wake. Amplitudes of components featuring frequencies attributed to the thermoacoustic effect are higher for points located in the blade wake. Only the strongest components are perceivable at the blade channel outlet.

Points Located at the Rotor Extension

For points located at the rotor extension, the distribution of amplitudes of the p' components depending on the channel height (Figure 10 and Figure 11) is the same as after the rotor. However, this distribution got partially blurred and at the channel extension the ratio of the p' amplitudes at 20% and 80% of the channel relative height does not exceed 5. The impact of the seal with its characteristic frequency (~3.9 kHz) can be observed at all points located at the channel extension. At the same time it should be noted that components featuring high frequencies are still present in the rotor extension, which probably results from the thermoacoustic effect.

CONCLUSIONS

This paper presents an analysis of the unsteady flow field in a turbine stage operating under a high load. The flow field unsteadiness was caused by a change in the mutual positioning of the stator and rotor blades and forced by aeroacoustic and thermoacoustic effects. The most common numerical method applied in fluid mechanics nowadays was employed for the calculations—the uRANS method available in the Ansys CFX software package. The analysis used the Adaptive-Scale Simulation (SAS) turbulence model, which belonged to the group of hybrid models. The main part of this paper is devoted to the analysis of the generation and propagation of pressure waves in a turbine stage. Comparing the results of the FFT analyses carried out for pressure fluctuations at selected points of the computational area, it is possible to make the following statements:

- The frequencies of the main harmonic components obtained from numerical computations subjected to the FFT analysis coincide with the frequencies of transition components resulting from theoretical dependences.

- The SAS model makes it possible to capture a large number of broadband noise components. These components are present especially in the range of low frequencies (up to 10 kHz), which is the acoustic range at the same time. Pressure fluctuations in the frequency range mentioned above result from the presence of large turbulence scales causing fluctuations in flow parameters.

- The SAS model also made it possible to capture noise components with relatively high amplitudes in the range of high frequencies. It is suspected that these components may result from the thermoacoustic effect induced by a high difference between the temperature of the blade wake right after the trailing edge and the temperature of the medium in its vicinity.

- The stator/rotor interface (keeping constant position with respect to the meshes) most probably did not interfere with the transmission of data concerning acoustic disturbances because significant and characteristic components generated in the seal area were also identified in the blade channel after the seal.

ACKNOWLEDGEMENTS

The authors would like to thank the Polish Ministry of Science and Higher Education for the financial support of the research project UMO-2011/01/B/ST8/03488.

REFERENCES

1.   Denton, J.D. (1993) Loss Mechanisms in Turbomachines. Journal of Turbomachinery, 115, 621-656. http://dx.doi. org/10.1115/1.2929299

2.   Dykas, S., Wróblewski, W. and Łukowicz, H. (2007) Prediction of Losses in the Flow through the Last Stage of Low-Pressure Steam Turbine. International Journal for Numerical Methods in Fluids, 53, 933-945. http://dx.doi.org/10.1002/fld.1313

3.   Dykas, S., Wróblewski, W., Rulik, S. and Chmielniak, T. (2010) Numerical Method for Modeling of Acoustic Waves Propagation. Archives of Acoustics, 35, 35-48.http://dx.doi. org/10.2478/v10168-010-0003-7

4.   Dykas, S., Wróblewski, W. and Machalica, D. (2013) Numerical Analysis of the Losses in Unsteady Flow through Turbine Stage. Open Journal of Fluid Dynamics, 3, 252-260. http://dx.doi.org/10.4236/ojfd.2013.34031

5.   Świrydczuk, J. (2009) Three-Dimensional Unsteady Interaction of Vortex Structures in a Steam Turbine Rotor. Proceeding of the 8th European Conference on Turbomachinery, Fluid Dynamics and Thermodynamics, Graz, 23-27 March 2009, 1197-1206.

6.   Lampart, P. (2009) Investigation of Endwall Flows and Losses in Axial Turbines, Part I. Formation of Endwall Flows and Losses. Journal of Theoretical and Applied Mechanics, 47, 321-342.

7.   Egorov, Y., Menter, F.R., Lechner, R. and Cokljat, D. (2010) The Scale-Adaptive Simulation Method for Unsteady Turbulent Flow Predictions. Part 2: Application to Complex Flows. Flow Turbulence Combust, 85, 139-165. http://dx.doi.org/10.1007/s10494-010-9265-4

8.   Menter, F.R. and Egorov, Y. (2010) The Scale-Adaptive Simulation Method for Unsteady Turbulent Flow Predictions. Part 1: Theory and Model Description. Flow Turbulence Combust, 85, 113-138. http://dx.doi.org/10.1007/s10494-010-9264-5

9.   Moffit, T.P., Szanca, E.M. and Whitney, W.J. (1980) Design and Cold-Air Test of Single-Stage Uncooled Core Turbine with High Work Output, TP-1680, NASA.

10.   Whitney, W.J., Stabe, R.G. and Moffitt, T.P. (1980) Description of the Warm Core Turbine Facility and the Warm Annular Cascade Facility Recently Installed at NASA Lewis Research Center, Technical Report, NASA. Technical Memorandum 81562.

Chapter 2

The Mechanical Impact of Aerodynamic Stall on Tunnel Ventilation Fans

A. G. Sheard[1] and A. Corsini[2]

[1]Fläkt Woods Ltd., Axial Way, Colchester CO4 5ZD, UK
[2]Dipartimento di Ingegneria Meccanica e Aerospaziale, Sapienza University of Rome, Via Eudossiana 18, Rome 00184, Italy

ABSTRACT

This paper describes work aimed at establishing the ability of a tunnel ventilation fan to operate without risk of mechanical failure in the event of aerodynamic stall. The research establishes the aerodynamic characteristics of a typical tunnel ventilation fan when operated in both stable and stalled aerodynamic conditions, with and without an anti-stall stabilisation ring, with and without a "nonstalling" blade angle and at full, half, and one quarter design speed. It also measures the fan's peak stress, thus facilitating an analysis of the implications of the experimental results for mechanical design methodology. The paper concludes by presenting three different strategies for tunnel ventilation fan selection in applications where the selected fan will most likely stall. The first strategy selects a fan with a low-blade angle that is nonstalling. The second strategy selects a fan with a high-pressure developing capability. The third strategy selects a fan with a fitted stabilisation ring. Tunnel ventilation system designers each have their favoured fan selection strategy. However, all three strategies can produce system designs within which a tunnel ventilation fan performs reliably in-service. The paper considers the advantages and disadvantages of each selection strategy and considered the strengths and weaknesses of each.

INTRODUCTION

The operating maps of fans and compressors are limited by the occurrence of aerodynamic instabilities when throttling the flow rate. Aerodynamic flow instabilities place considerable mechanical stress on the rotors, which can eventually lead to mechanical failure. Rippl [1] conducted strain gauge measurements on axial compressors, concluding that alternating stress in vanes exceeding stable operation by a factor of five under "rotating stall" conditions. This leads to rapid fatigue failure of the blades. In contrast, a "surge" can lead to the heightening magnitude of bending stress enough to cause a mechanical failure during the surge event itself.

Fan designers classically produce a mechanical design that can withstand the alternating loads imposed on the fan blades associated with rotating stall, and therefore mechanical failure during a stall event is not instantaneous. Aluminium is both low cost and light weight, and consequently the fan designers' preferred choice of blade material. A weakness of aluminium as a structural material is its propensity to fail in fatigue. As such, fan blades that do not typically instantaneously fail during rotating stall fail in fatigue sometime later. The latter failure occurs as a consequence of a fatigue-induced crack initiated in a blade as a consequence of the higher stress during the rotating stall that then goes on to propagate during stable operation.

This paper studies the impact of rotating stall, generally referred to as "aerodynamic stall" within the fan industry, on the mechanical performance of a typical tunnel ventilation fan. The paper starts with a brief literature review relating to fan, blower, and compressor aerodynamic stall before moving on to review the antistall concepts that other scholars have developed in their attempts to improve axial decelerating turbomachinery aerodynamic stability. Placing strain gauges in the location of the fan blades' peak stress, the authors were able to establish the mechanical impact of aerodynamic stall with and without an antistall stabilisation ring, with and without a "nonstalling" blade angle and at full, half, and one quarter design speed.

The chosen test matrix incorporated those fan configurations typically utilised in tunnel ventilation applications. The purpose is to establish the increase in peak blade stress associated with transition from stable and stalled aerodynamic conditions. The objective of establishing the increase is to facilitate an analysis of its mechanical consequence. The outcome of the reported research is the identification of change in mechanical safety factors associated with a fan being driven into aerodynamic stall. To the best of the author's knowledge, the results presented in this paper are the first time the effectiveness of an antistall ring has been reported in the literature. The reported research has established that an antistall ring provides some mechanical protection in the event of aerodynamic stall, but not complete protection. The reported research demonstrates the error in the wide-spread assumption amongst industrial fan designers that antistall rings provide complete mechanical protection in the event of aerodynamic stall.

The paper concludes with an analysis of the significance of the results for fan design praxis, recommendations on fan selection strategy and mechanical design methodology for those tunnel ventilation fans applied in selected applications where the fan will most likely stall.

AERODYNAMIC STALL

Scholars have examined the detection and analysis of different forms of aerodynamic instability since the 1950s. According to Gravdahl and Egeland [3], two main types of aerodynamic flow instability exist in compressors: (i) "rotating stall" (in which regions of reversed flow occur locally) and (ii) "surge" (which is characterised by periodic backflow over the entire annulus involving violent oscillations in the compression system).

The first of these, "rotating stall," is a mechanism by which the rotor adapts to a reduction in flow rate, which results in circumferentially nonuniform flow patterns rotating in the annulus. Researchers have studied the problem of rotating stall in axial flow compressors in multistage machines [4–6]. The earlier work of Emmons et al. [7] was one of the first attempts to describe the mechanism underlying the propagation of rotating stall. In reviewing the evolution of rotating stall, Cumpsty [8] noted that a drop in overall performance can occur as either a "progressive stall" or an "abrupt stall." Engineers usually associate the former with a part-span stall which results in a small performance reduction; whereas, they associate the later with a full-span stall and a large reduction in performance. Notably, the partspan rotating stall typically occurs in single blade rows [8] and usually leads to more complex disturbances in single rotor or stage machines than in multistage compressors [6].

The fan under scrutiny in the reported research is a typical example from a family of tunnel ventilation fans and has been the subject of recent experimental investigation [9–11]. The investigation focused on the stall modes, identifying a rotating stall inception mechanism driven by circumferentially localised pressure disturbances confined to the blade passage's tip region. Localisation of the disturbances in the blade tip region supports the hypothesis of a causal link between tip clearance flow and stall inception.

Some scholars have focused on the physics underlying the tip clearance flow related mechanisms that can lead to the formation of pressure disturbances. Among them, Koch and Smith [12] and Saathoff and Stark [13] have observed experimentally that a fan reaches the limit of its pressure developing capability when the interface between the incoming flow and tip clearance vortex region lined-up with the leading edge plane at the blade tip. Numerical simulation has revealed two more mechanisms related to the onset of tip blockage growth [14,15], namely: (i) the backflow of tip clearance fluid at the trailing edge impinging on the blade pressure side of adjacent blades and (ii) the spillage of tip clearance fluid ahead of the blade leading edge below the blade tip into the next blade passage.

Scholars who have studied aerodynamic instability in fans, and compressors have suggested that some features of the tip flow of both are directly responsible for the generation of short wavelength disturbances (also called "spikes" or "pips") that cause the inception of localised part-span stall cells [16–18].

HISTORICAL OVERVIEW OF ANTISTALL CONCEPTS

Given the potentially catastrophic consequences of a stall event, there is an incentive for developing technologies that can extend the stable operating range of axially decelerating turbomachinery without undue performance degradation. Previously, Hathaway [19] systematically reviewed techniques and design concepts to improve the stall-free operating margin or to suppress a stall event.

Hathaway noted the earliest proposed techniques from the 1950s that had successfully extended the axial compressor's stable operating range: Wilde [20], on behalf of Rolls-Royce Ltd, and Turner [21] on behalf of Power Jets Ltd, filed patents. These concepts were both based on the treatment of the casing end-walls motivated by a desire to control the boundary layer development by combining rear air bleeding and front reinjection [20] or the use of holes and slots as a method of promoting turbulence, and in so doing, energising the end-wall flow [21].

Griffin and Smith [22] conducted the first systematic experimental campaign on so-called "porous end-walls" in compressor rotors during the 1960s at NASA. Their work demonstrated improved stall margins in cascade tunnels irrespective to the air blowing/bleeding. In a similar vein, scholars studied casing treatments, specifically holes, slots, and grooves with and without plena in the 1970s [23–25]. They found their effectiveness primarily associated with delaying the onset of stall in tip-limited blade rows.

Takata and Tsukuda [26] conducted detailed measurements within casing slots and found that they achieved their antistall effect as a consequence of periodic pumping of the flow within the slots into the main stream. Moreover, Greitzer et al.'s [27] investigations demonstrated that the end-wall treatments were effective mostly in high-solidity blade rows prone to wall stall, but not in low-solidity rotor affected by blade stall.

In the 1990s, researchers proposed concepts to exploit the potential benefits of flow bleeding (from the stalled region) and blowing (into the clean inflow), and in so doing, revisited earlier stall inception concepts. Most notably, Koff et al. [28], Khalid [29], Nolcheff [30], and Gelmedov et al. [31] patented different variants of recirculating casing treatment. A common theme with the different recirculating casing treatments was the provision of a path through the casing for low momentum fluid recirculating upstream from the blade tip leading edge.

More recently, Hathaway [19] observed that the most significant advances in antistall devices have resulted as a consequence of insight into the flow mechanism that researchers associate with three specific technologies: first, circumferential grooves; second, tip injection control technology; third, stage recirculation devices. Fan and blower designers most favour the stage recirculation devices.

PASSIVE CONTROL BY STABILISA-TION RINGS

Since the early 1960s, scholars have endeavoured to develop stage recirculation devices tailored to the pressure rise and volume flow rate ranges typical of industrial fans. Ivanov [2] received the first patent. The concept is of an annular "slit" in the casing upstream of the blades (Figure 1) that stabilises fan performance as it approaches stall.

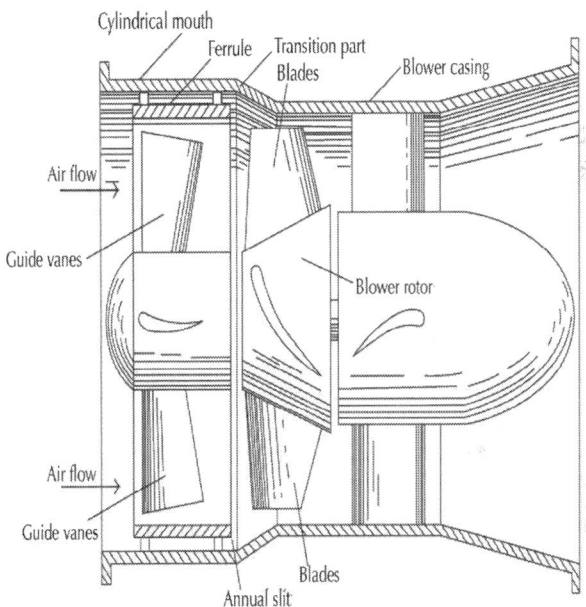

Figure 1: The proposed "blower arrangement" [2].

The slit enabled air to recirculate through the slit formed between the cylindrical mouth and ferrule, Figure 1. A set of guide vanes then redirected the recirculating flow in an axial direction as it turns back and reenters the fan blades. As the fan approaches stall, the slit and static vanes provide a path for low momentum flow to recirculate. In practice, this stabilises fan performance.

Karlsson and Holmkvist [32] filed a patent application on 15 March 1984, developing and enhancing Ivanov's [2] patent by incorporating static vanes into the casing. Then, Bard [33] named the vanes embedded within the fan casing a "stabilisation ring." Miyake and Inaba [34] further developed and patented the original concept proposing the use of air-separators based on an open circumferential cavity facing the rotor-blade tips, which Yamaguchi et al. [35] further developed. Similarly, Kang et al. [36] optimised casing recess geometries and their relative position to the blade rows.

Despite subsequent developments to the concept, the fan community has mostly adopted Karlsson and Holmkvist's [32] configuration. In practice, the concept has proven highly effective as with the stabilisation ring guide, vanes remove the momentum component both radially and circumferentially and reinject the flow in the axial direction. The flow through the stabilisation ring vanes is turned such that it exits the vanes upstream from the impeller, reenergised, and flowing in an axial direction.

The effect of the stabilisation ring on the fan characteristic is to eliminate the sharp drop in its pressure developing capability, which engineers classically associate with fan stall. The primary characteristic of a tunnel ventilation fan fitted with a stabilisation ring is continuously rising pressure back to zero flow. It was this modification in the fan characteristic that led to tunnel ventilation fan designers widely embracing the use of stabilisation rings.

A continuously rising characteristic facilitates multiple fan operation in parallel. As a fan speed falls, its pressure developing capability also falls. During a fan's starting and stopping transient, its pressure developing capability will be below what other fans generate when operating in parallel. As a consequence, a fan in parallel operation will inevitably drive transiently into stall each time it starts or stops. During the 1980s, variable speed drives were not widely available. Therefore, varying the speed of all fans in a parallel installation was not practical, making it inevitable that individual fans would have to start and stop, whilst others ran at full speed.

The ability of the stabilisation ring to facilitate the starting and stopping of individual fans when in parallel operation was critically important. Application of the stabilisation ring largely eliminated inservice mechanical failure in tunnel ventilation fan parallel operation.

A particular feature of the environment within which tunnel ventilation fans operate is the pressure pulses associated with the movement of a train through a tunnel. Pressure pulses can be up to ±50% of the overall work coefficient. Such pressure pulses drive the fan first up and then down its characteristic operating range, Figure 2. To ensure that the tunnel ventilation fan continues to operate in an aerodynamically stable manner during this pressure transient, aerodynamic design of the fan requires the incorporation of sufficient margin to ensure that the fan does not stall due to high positive or negative inlet flow angle.

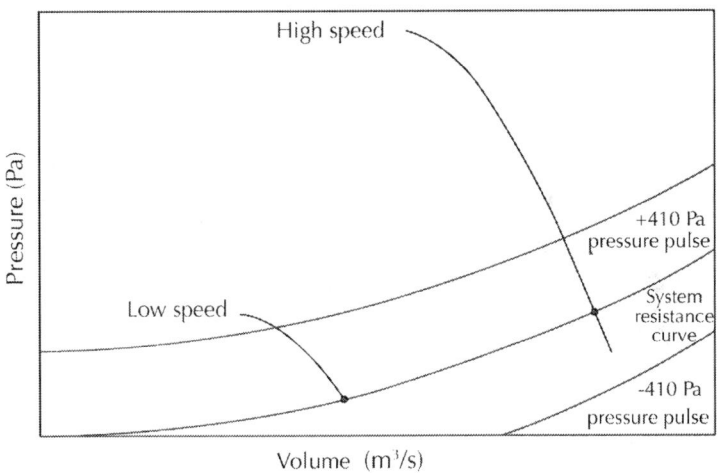

Figure 2: Fan performance under a pressure pulse at full and part speed.

This propensity to stall under large pressure fluctuations is complicated in offdesign conditions when a tunnel ventilation fan operates at partial speeds. When a fan operates at reduced speed, its flow and pressure-developing capability also reduces. Because the pressure pulse of a passing train remains constant, there will be a critical speed when the fan is close to stall, but has not actually stalled. Below that critical speed, the fan stalls in positive incidence as the train approaches and then stalls in negative incidence as the train departs, Figure 2. This combination of positive-incidence aerodynamic stall and negative-incidence aerodynamic stall causes a significant increase in the unsteady forces applied to the fan blades.

TECHNOLOGY DESCRIPTION

In order to establish the likelihood of mechanical failure of tunnel ventilation fans in applications where aerodynamic stall is likely due to the presence of pressure pulses, the researchers selected a fan typical of tunnel ventilation applications as Table 1 below illustrates.

Table 1: Fan data

Nominal speed	980 rpm
Tip speed	115 m/s
Nominal pressure coefficient, Ψ_{nom}	0.189
Nominal flow coefficient, Φ_{nom}	0.220
Duty point efficiency, η_{tot}	0.69
Tip diameter	2,240 mm
Blade height	720 mm
Blade chord at the tip	163 mm
Tip stagger angle	70°
Tip gap (% of fan diameter)	0.45%
Blade count	16
Tip solidity	0.37

The chosen fan was from a family of tunnel ventilation fans. Although each fan within the family is physically different, all are designed using the same mechanical design methodology. The researchers chose the fan configurations for experimental testing as a consequence of the application into which manufacturers supply tunnel ventilation fans. Tunnel ventilation fans are typically subjected to pressure pulses. Therefore, the researchers tested the fan

- with a stalling blade angle, without a fitted stabilisation ring at design speed
- with a stalling blade angle, with a fitted stabilisation ring at design speed
- with a nonstalling blade angle and no stabilisation ring at design speed
- with a stalling blade angle and no stabilisation ring at 50% design speed;(v)with a stalling blade angle and no stabilisation ring at 25% design speed.

The researchers installed the fan in a ducted test system. They measured the fan performance according to the International Standard ISO 5801:2007 (2007). By throttling flow downstream from the fan rotor, the researchers induced aerodynamic instabilities of interest. During the flow/pressure throttling, the fan remained in rotating stall without going into surge, irrespective of the rotor speed. The rotor aerodynamic load and the plenum geometry ensured that the system could not develop a counter-pressure high enough to induce a surge.

The researchers developed a solid model of the test fan's blade using a CAD package, and they used a finite element analysis programme for structural analysis. They applied centrifugal force and bending moments (due to the design radial work distribution) using Sheard et al.'s [37] original method using nodal forces in the finite element analysis boundary conditions in order to calculate blade stress. The authors applied strain gauges to three blades in the three locations that they predicted as the blades' high-stress regions.

Stain gauges were applied using the method of Boyes (2003 : 77). After chemical cleaning of the blades to be instrumented, foil gauges were applied. Foil gauges were chosen as their flatness makes adhesion easier and improves heat dissipation. The instrumentation system utilised was an adaption of that originally developed by Wasserbauer et al. [38] who reported the design of a low-speed axial compressor test facility at what was then named NASA Lewis. Strain gauge leads are routed from the rotating to static frame of reference though a four-channel electrical slip ring to a set of strain gauge amplifiers. The output of the strain gauge amplifiers was then logged using a PC-based data acquisition system running the software package LabView.

EXPERIMENTAL RESULTS

Application of multiple strain gauges to separate blades enabled the authors to experimentally determine the actual highest stress location, as well as the impact of manufacturing tolerances from blade to blade. The variation in strain gauge output from blade to blade at nominally the same location on different blades was 2-3%. This variation constitutes a combination of errors associated with gauge calibration, uncertainty in gauge location, and blade-to-blade variation of blade geometry.

Using data from a typical strain gauge located at the highest stress position on one blade, the researchers established fan performance with and without a fitted stabilisation ring as Figure 3 illustrates. When throttling the fan without a fitted stabilisation ring, pressure rises until it reaches a peak and then falls as the fan stalls. This is the classical fan characteristic. In the study, the blade's peak alternating stress increased from 2.27 MPa (Point A, Figure 3) to 3.53 MPa as the researchers throttled the fan. As the fan stalled, peak alternating stress increased to 16.00 MPa (Point B, Figure 3).

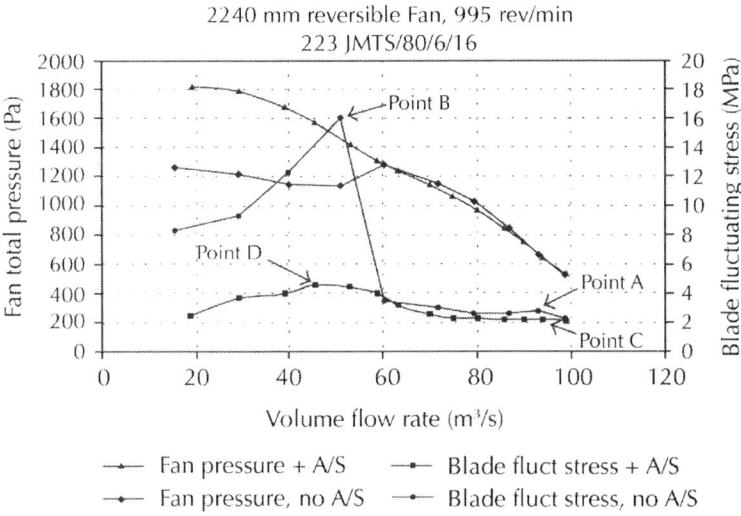

Figure 3: Stall characteristics of the test fan with and without a fitted stabilisation ring.

When the researchers throttled the fan with a stabilisation ring, pressure rose continuously with no evidence of a reduction in pressure developing capability as the fan passed through the point at which it stalled without a stabilisation ring, Figure 3. The fan characteristic is remarkable in that the pressure rises so smoothly that it is barely possible to identify the onset of stall from the fan's flow/pressure characteristic. However, in studying the strain gauge data, it is apparent that the initial alternating stress level is 2.13 MPa (Point C, Figure 3) and remains lower than a fan without a stabilisation ring until the onset of stall. As the fan fitted with a stabilisation ring approaches stall, there is a single point (at 60 m³/s) where the data from the fan without a stabilisation ring measures lower stress than the fan with a stabilisation ring. Alternating stress in the fan with a stabilisation ring goes on to peak at 4.60 MPa (Point D, Figure 3).

Using data from a typical strain gauge located at the highest stress position on one blade, the researchers established fan performance with a nonstalling blade angle and no fitted stabilisation ring. The researchers compared data with data for a stalling blade angle with a fitted stabilisation ring as Figure 4 illustrates. The authors compared the two data sets as a nonstalling blade angle without a stabilisation ring or stalling blade angle with the stabilisation ring representing the two available choices to tunnel system designers who wish to specify a "stall tolerant fan."

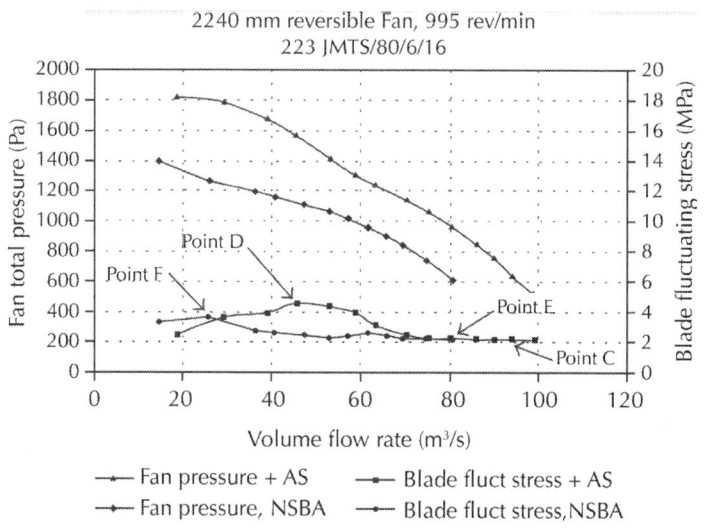

Figure 4: Stall characteristics of the test fan with a nonstalling blade stagger angle.

The change in blade angle from stalling to a nonstalling reduces the fan pressure developing capability by approximately 25%, Figure 4. The researchers tested the fan with the stalling blade angle and fitted stabilisation ring under stable and stalled aerodynamic conditions. The alternating stress increased from the initial alternating stress level of 2.13 MPa (Point C, Figure 4), going on to peak at 4.60 MPa (Point D, Figure 4). In contrast, a nonstalling blade angle without a stabilisation ring, increased from an initial 2.19 MPa (Point E, Figure 4), going on to peak at 3.68 MPa (Point F, Figure 4).

The peak alternating stress with the nonstalling blade angle and no fitted stabilisation ring was 20% lower than a stalling blade angle and fitted stabilisation ring. The researchers considered the uncertainty of the measurement approximately 2%, and consequently, a 20% reduction in peak alternating stress is an order of magnitude greater than the uncertainty of the measurement. The above result indicates that a fan with a nonstalling blade angle and no fitted stabilisation ring will be subject to lower peak alternating stress during aerodynamic stall than the same fan with stalling blade angle and a fitted stabilisation ring.

Using data from a typical strain gauge located at the highest stress position on one blade, the researchers established fan performance with a stalling blade angle, and no stabilisation rig fitted at 100%, 50%, and 25% design speed. The availability of variable speed drives has resulted in tunnel ventilation fans routinely operating at part speed. The researchers compared the data at 100% speed with data at 50% speed, as Figure 5 illustrates. The authors compared the two data sets, as tunnel ventilation fans selected to not stall in the presence of a pressure pulse at 100% speed, routinely stall in the presence of the same pressure pulse at 50% speed.

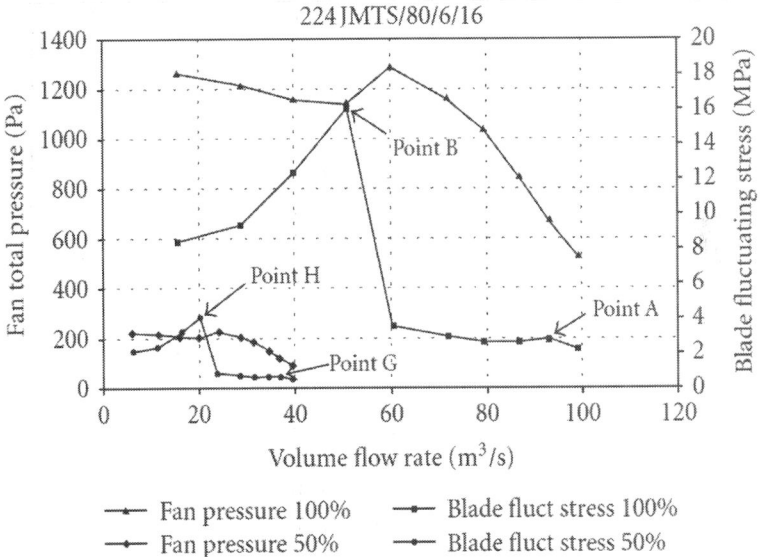

Figure 5: Stall characteristics of the test fan with a stalling blade stagger angle at full and half speed.

The same alternating stress data for a fan with stalling blade angle and no fitted stabilisation ring is plotted, with alternating stress increasing from the initial alternating stress level of 2.27 MPa (Point A, Figure 5), going on to peak at 16.00 MPa (Point B, Figure 5). In contrast, the same fan running at 50% speed had an initial alternating stress of 0.57 MPa (Point G, Figure 5), going on to peak at 4.00 MPa (Point H, Figure 5). As this paper previously mentioned, at full speed, the alternating stress increased to 3.53 MPa immediately prior to aerodynamic stall. As such a fan running at a speed of 100% (with no aerodynamic stall in the event that a pressure pulse) would see a maximum alternating stress of 3.53 MPa. The same fan running at 50% speed has a reduced pressure developing capability and so would in all probability stall in the presence of the same pressure pulse, and in doing so, be exposed to a peak alternating stress of 4.00 MPa.

In some metro tunnel ventilation systems, it is customary to operate tunnel ventilation fans at 25% design speed. The same alternating stress data for a fan with stalling blade angle and no fitted stabilisation ring is plotted for the fan running at 50% speed, with alternating stress increasing from the initial alternating stress level of 0.57 MPa (Point G, Figure 6), going on to peak at 4.00 MPa (Point H, Figure 6). In contrast, the same fan running at 25% speed had an initial alternating stress of 0.14 MPa (Point I, Figure 6), going on to peak at 1.00 MPa (Point J, Figure 6). The alternating stress level during aerodynamic stall at 25% speed (1.00 MPa) is significantly lower than the alternating stress level during stable operation at 100% speed (3.53 MPa) and, therefore, the researchers concluded that it posed no risk to the fan's mechanical integrity.

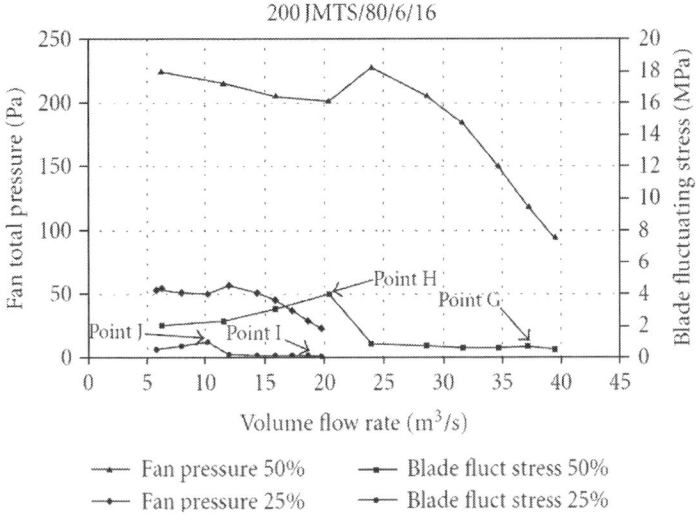

Figure 6: Stall characteristics of the test fan with a stalling blade stagger angle at half and quarter speed.

STRUCTURAL ANALYSIS

The term "fatigue" refers to the phenomenon whereby virtually all materials will break under numerous stress repetitions that are not sufficient to produce an immediate rupture in the first instance. In this regard, fan blades are subject to fatigue stress induced by (i) the mean force arising from rotation and aerodynamic loading and (ii) the alternating force produced by variations in lift as the fan rotates. The combination of mean and alternating blade forces result in mean and alternating blade stress. This makes the blades inherently susceptible to fatigue.

The endurance limit corresponding to any given range of stress variation has been the subject of extensive study, reviewed by amongst others Young [39], as has the ability of a wide range of materials to withstand different combinations of mean and alternating stress. Manufacturers usually produce tunnel ventilation fan blades from aluminium and the ability of aluminium to resist fatigue for a fixed alternating stress reduces as mean stress increases. When researchers study material test data for various levels of mean and alternating stress, they derive a relationship known as the Gerber Line, Figure 7.

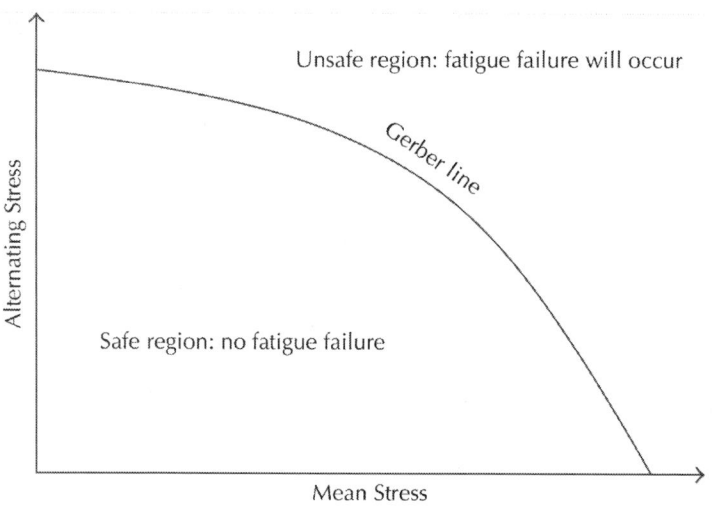

Figure 7: The curve of best fit through material test data is the Gerber Line.

Gerber [40] himself derived this line and proposed a parabolic relationship between alternating stress and mean stress in iron structures. The maximum alternating stress level σ for any mean stress in the material, up to the tensile strength of the material, is given by the expression:

$$\pm\sigma = \pm\sigma_0 \left\{ 1 - \left(\frac{\sigma_m}{\sigma_t} \right)^2 \right\}.$$

(1)

σ_0 is the alternating stress level that constitutes the fatigue limit of the material with zero mean stress. σ_t is the tensile strength of the material. σ_m is the mean stress in the material.

The ability of a given aluminium alloy to resist the effect of mean and alternating stress is dependent on the maximum defect size in the material samples. The larger the defect, the lower the level of mean and alternating stress required to induce fatigue failure. Tunnel ventilation fans manufacturers, therefore, first experimentally establishes the relationship between mean and alternating stress for a given defect size, and then undertakes X-ray examination of all rotating components to ensure that the maximum defect size is below that on which they established the Gerber Line.

If the peak mean/alternating stress point is below the Gerber Line, the fan blade should not fail due to fatigue. However, in practice, there is some uncertainty about the location of all Gerber Lines as they are derived from experimental data. Additionally, the ability to calculate mean and alternating stress levels is imperfect as a consequence of assumptions during the modelling process. Therefore, in practice, tunnel ventilation fan designers classically choose to design fans with a safety factor of two. In this context, we define a safety factor of two as an alternating stress half that of the Gerber line at a given value of direct stress.

The authors assessed the significance of the measured alternating stress results. The manufacturers designed the family of fans that the authors used in the reported research with direct and alternating stress levels that would fall on a Gerber Line calculated with a safety factor of two. The authors combined predicted direct and measured alternating stress levels for the fan operating on a stable part of its characteristic, with no fitted stabilisation ring and a stalling blade angle to give a safety factor of 2.3, Table 2. The experimentally derived safety factor was greater than two, giving confidence in the conservative nature of the manufacturers mechanical design methodology.

Table 2: Safety factor derived from strain gauge data for a fan at full speed with and without a fitted stabilisation ring

Fan type	% Full speed	Normal operation safety factor	Stalled operation safety factor
Plane casing, stalling blade angle	100	2.3	0.3

Antistall casing, stalling blade angle	100	2.5	1.1
Plane casing, nonstalling blade angle	100	2.4	1.5

Throttling the fan until stall without a stabilisation ring resulted in increasing alternating stress. The resultant combination of direct and alternating stress is significantly beyond the Gerber Line, giving a safety factor of 0.3 (Table 2). From this, we may conclude that if this fan operated in the stalled condition for an extended period of time, it would suffer a fatigue-related failure.

The researchers measured direct and alternating stress levels during stable and stall conditions for the test fan with stalling blade angle and a fitted stabilisation ring, Table 2. The reduction in alternating stress during stable operation resulted in the fan operating with a slightly higher safety factor of 2.5. In stalled operation, the alternating stress increased and, in so doing, reduced the safety factor to 1.1. As a safety factor of 1.1 is greater than one, the mechanical design of the tested fan can tolerate the increase in alternating stress. However, the uncertainty of the Gerber Line location is significant enough for a conservative tunnel ventilation fan designer to consider it low. The researchers measured direct and alternating stress levels during normal and stall conditions for the test fan with a nonstalling blade angle and no stabilisation ring, Table 2. The increase in alternating stress compared to the same fan with stalling blade angle and a fitted stabilisation ring during stable operation resulted in the fan operating with a slightly lower safety factor of 2.4. In stalled operation, the alternating stress increased and, in so doing, reduced the safety factor to 1.5. As a safety factor of 1.5 is greater than one, the mechanical design of the tested fan can tolerate the increase in alternating stress, and therefore, the risk of a fatigue-induced mechanical fan failure is low.

Operating the fan at 100%, 50%, and 25% speed in both stable and stalled operation facilitated in the calculation of six safety factors, Table 3. This study reports a tunnel ventilation fan safety factor during stable operation at 100% speed of 2.3 (0.3 higher than the desirable minimum). In a stalled condition, we reduced the safety factor to 0.3, which is significantly lower than 1.0. This indicates that if the fan continued to operate in stall at 100% speed, the blades would suffer fatigue failure.

Table 3: Safety factor derived from strain gauge data for a fan at full and part speed without a fitted stabilisation ring

Fan type	% Full speed	Normal operation safety factor	Stalled operation safety factor
Plane casing, stalling blade angle	100	2.3	0.3
Plane casing, stalling blade angle	50	10.0	2.5
Plane casing, stalling blade angle	25	106.0	7.3

During normal operation, safety increased from 2.3 at 100% speed to 10.0 at 50% speed, a factor of approximately four. Mean stress in a fan blade reduces with the square of speed, therefore reducing speed by half than expected, thus increasing the safety factor by four. When the authors operated the fan at 50% speed during stalled operation, the safety factor was 2.5, Table 3 (an increase of more than six compared to the same fan operating in stall at 100% speed). The increase in the safety factor is a consequence of the aerodynamically induced alternating stresses (when operating the fan in stall) falling more rapidly then the mean stress falls when the authors reduced fan speed from 100% to 50%.

The safety factor at 50% speed when operating in stall (2.5) was slightly higher than the safety factor at 100% speed in stable operation (2.3). As the safety factor at 50% when operating in stall is higher then the safety factor of the same fan at 100% speed in stable operation, the authors observed that the fan was less likely to fail mechanically at 50% speed then when at 100% speed in stable operation.

The above result is significant because tunnel ventilation fans in metro applications routinely operate at both 50% and 25% speeds. As a consequence of the reduced pressure-developing capability of the fans at reduced speed, fans in metro applications are routinely driven into stall. The calculated safety factors at 25% design speed are so high during both stable and stalled operation that we can conclude that aerodynamic stall poses no threat to the mechanical integrity of the tunnel ventilation fan.

The above results indicate that the tested fan may operate at 100% speed in the stable part of its characteristic, with a mechanical safety factor of 2.3. The same fan can operate also at 50% speed in aerodynamic stall with a mechanical safety factor of 2.5. When the authors scale the fan characteristic to 90% speed, the pressure-developing capability reduced to the point where a 500 Pa pressure pulse would take the fan within 5% of the fan's peak pressure-developing capability.

Next, the authors scaled the direct stress from 50% to 55% design speed and recalculated the associated mechanical safety factor with operation in an aerodynamically stalled condition. The mechanical safety factor reduced from 2.5 at 50% design speed to 2.0 at 55% design speed. From the above, the authors concluded that the tested fan could operate at up to 55% design speed in aerodynamic stall and down to 90% design speed without stalling. As such, the authors observed that if they fitted this particular fan with a variable speed drive (VFD), the "forbidden" speed range should be 55–90% speed for the tested design point, assuming a 500 Pa pressure pulse.

STRATEGY FOR FAN SELECTION

The demand for new mass-transit systems generally, and metro systems in urban areas specifically, has increased rapidly over the last two decades and continues today. However, recent changes in market requirements for tunnel ventilation fans present fan designers with a challenge. The proposed tunnels for the next generation of mass transit systems are longer than the historic norm, and the trains that run in them are to run faster. These two factors result in mass-transit systems requiring higher pressure ventilation fans (as the tunnels are longer) with the capability of operating under the influence of larger pressure pulses (as the trains are running faster).

An additional factor increasing the magnitude of pressure pulses in metro systems is the trend towards the use of platform screen doors. Platform screen doors at metro stations screen the platform from the train. They are a relatively new addition to metro systems and are today in wide use in Asia and Europe. Passenger safety is driving the adoption of platform screen doors. By separating the platform from the train, platform screen doors prevent the travelling public either accidentally or deliberately falling into the path of oncoming trains. Additionally, metros designed to use driver-less trains are only considered safe if platform screen doors are included in the design. Consequentially, it is likely that an increasing proportion of new and refurbished metro stations will include platform screen doors.

Historically, a pressure of 1,200 Pa with pressure pulses of typically 300 Pa has been typical in tunnel ventilation system application. Today, a pressure of 1.500 Pa with pressure pulses of 500 Pa is typical. Increasing pressure and pressure pulse size increase the importance of tunnel ventilation fan selections that either avoid or manage the effect of fan stall.

A modern mass transit system in a busy urban area can have 500 trains a day passing each tunnel ventilation shaft. Consequently, the fans in those ventilation shafts are subjected to 500 pressure pulses a day and therefore, are potentially driven into stall each time. With a typical inservice life of 20 years, the probability of incorrectly selected fans for the application suffering a fatigue-induced mechanical failure becomes high.

Tunnel ventilation fan designers have classically utilised one of three approaches during the selection of tunnel ventilation fans that must operate in the presence of pressure pulses.

- Select a fan with a nonstalling blade angle, such that as the fan is driven out of its normal operating range, mechanical stress increases within manageable limits.
- Select a fan with a high enough pressure-developing capability to operate with a pressure pulse without stalling.
- Select a fan with a stabilisation ring, such that as the fan is driven out of its normal operating range, mechanical stress increases within the limit of the mechanical design.

All three fan selection strategies are valid, and tunnel ventilation system designers have used each for tunnel ventilation system design. The first strategy, a fan with a nonstalling blade angle is the most conservative selection strategy. One may select the fan close to its optimum operating point, without having to compromise the selection to accommodate a pressure pulse within the stable operating range.

The second strategy, a fan with a high enough pressure-developing capability to operate with a pressure pulse without stalling, works well with smaller pressure pulses. However, as the size of the pressure pulse increases, the fan's operating point moves further from the optimum resulting in a less efficient fan selection. Despite the reservation about fan efficiency, this strategy avoids the fan stalling completely in the event of foreseen events. Tunnel ventilation system designers who are confident in their ability to predict the conditions under which the fans will operate during all tunnel ventilation system scenarios favour this second strategy.

The third strategy, a fan with a fitted stabilisation ring, works well with larger pressure pulses, allowing fan selection close to its peak aerodynamic efficiency and then effectively managing the mechanical consequences of driving into stall under the influence of pressure pulses. Tunnel ventilation system designers who favour this strategy argue that it is not possible to guarantee that the fan will never drive into stall; therefore, a stabilisation ring that provides mechanical protection in the event of aerodynamic stall is prudent. The tunnel ventilation system designers who favour this approach cite the possibility that occasionally two trains might pass close to a ventilation shaft, effectively doubling the size of the pressure pulse.

To facilitate comparison of the three strategies, the authors have made assumptions typical of a present day urban area metro system as Table 4 illustrates.

Table 4: Factors impacting on fan capital and through life cost

Design point pressure	1.500 Pa
Pressure pulse	500 Pa
Design point flow	85 m³/s
Fan type	Reversible, 300°C for 2 hours
Running hours per year	4.400 (12 hours a day)
Cost of electricity	0.04 £ per kW/Hour
Cost of capital	8%
Period of assessment	10 years

When considering the required pressure, flow and size of the pressure pulse, the first strategy, a fan with a nonstalling blade stagger angle, results in a fan of 2.50 metre diameter with a design point efficiency of 71% (see Figure 8). The second strategy, a fan that can accommodate the pressure pulse, results in a fan of 1.80 metre diameter with a design point efficiency of 66% (see Figure 9). The third strategy, use of a stabilisation ring, results in a fan of 2.24 metre diameter with a design point efficiency of 69% (see Figure 10).

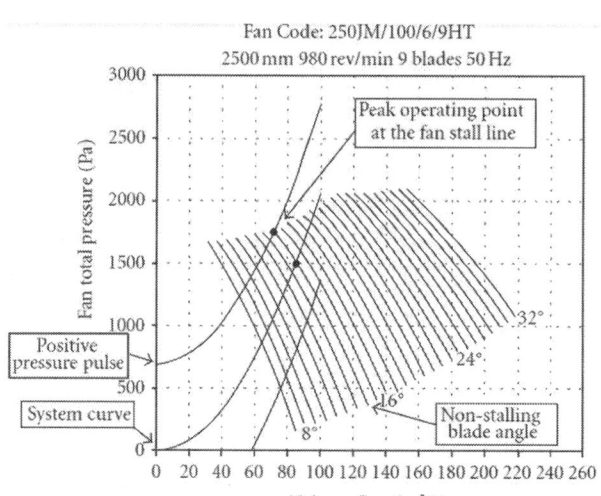

Figure 8: Optimum fan selections for a common duty point, fan selection strategy one: nonstalling blade stagger angle.

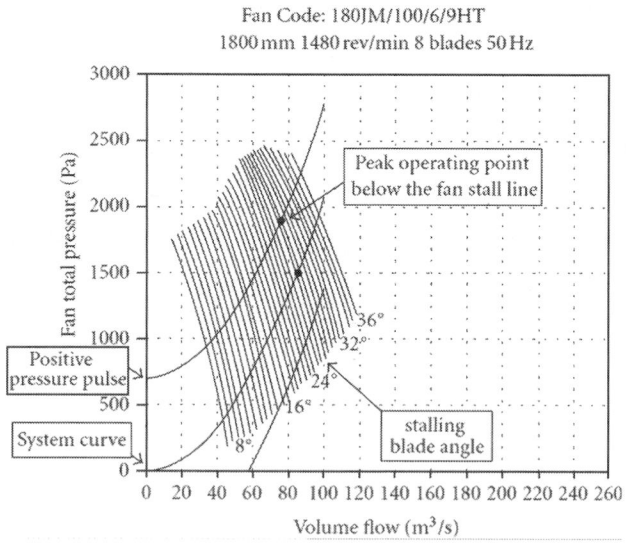

Figure 9: Optimum fan selections for a common duty point, fan selection strategy two: high-pressure capability.

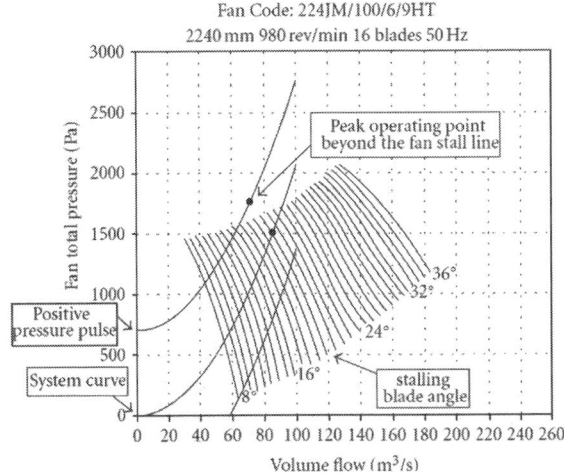

Figure 10: Optimum fan selections for a common duty point, fan selection strategy three: stabilisation ring.

We can combine the capital cost of each fan selection strategy with the Table 4 assumptions to calculate throughlife cost of each strategy (See Table 5). We define the initial capital cost in Table 5 as the initial cost of purchasing and installing the ventilation fan. Table 5 defines operating cost as the cost of fan purchase and installation plus electricity costs over ten years, based on Table 4 assumptions. To facilitate a direct comparison between initial capital cost and operating cost, Table 5 presents the net present value of operating costs.

Table 5: Capital cost and ten year though life cost of each selection strategy

	Fan diameter (m)	Fan efficiency	Fan investment cost (£)	Motor power (kW)	Electricity cost/year (£)	Electricity cost (10 yrs) (£)	Total fan and running costs (£)	Running cost as a percentage of total cost
Strategy 1	2.5	71%	28.500	185	32.412	217.488	245.988	88%
Strategy 2	1.8	66%	23.000	214	37.493	251.582	274.982	92%

Strategy 3	2.24	69%	32.000	190	33.288	223.366	255.366	87%

As previously mentioned, different tunnel ventilation system designers favour different fan selection strategies. However, all three strategies have reliable inservice records. In this example, the largest fan (Strategy One) has the lowest operating cost over ten years, Table 5, despite not having the highest initial cost, making this selection strategy attractive to those seeking the lowest cost of ownership.

In this example, the smallest fan (Strategy Two) has the highest operating cost over ten years, Table 5, despite having the lowest initial cost, making this selection strategy apparently unattractive to those seeking lowest cost of ownership. However, if one accounts for the cost of excavating an underground plant room, the cost of building a plant room for a 2.5 metre diameter fan (Strategy One) may be significantly higher than the cost of building a plant room for a 1.8-metre-diameter fan (Strategy Two), therefore making the second selection strategy attractive.

In this example, the medium size fan (Strategy Three) has the highest initial cost, Table 5, reflecting the manufacturer's cost of a fan casing with a stabilisation ring. Operating costs fall between costs for Strategy One and Two, reflecting the efficiency of the medium-sized fan that is between those of the largest and smallest.

Stabilisation rings can reduce the fan's efficiency, due to the recirculating flow in the blade tip region. For fans intended for use at ambient temperature only, the reduction in efficiency can be significant. In this study, the authors have assumed that the fans associated with each strategy are designed in accordance with the requirements of EN 12101-3 (2002) and ISO 21927-3 (2006) for once-only emergency operation at 300°C. The use of aluminium blade results in the blades expanding more rapidly then the steel casing as temperature rises. Consequently, the blade tip to casing clearance has to increase at ambient temperature to prevent it closing at high temperature. The increase in blade tip to casing gap from typically 0.25% of fan diameter (for ambient duty only) to 0.45% (for once-only 300°C emergency duty) reduces fan efficiency. In practice, the stabilisation ring does not reduce efficiency further when a fan has a large tip gap to facilitate high-temperature operation. Consequently, the fan that the authors used in this study confirms the general rule of thumb that larger fans are more efficient for a fixed duty point.

SUMMARY AND CONCLUSIONS

The experimental results are significant in that they provide insight into a likely reason for tunnel ventilation fans' inservice failure. In this study, the alternating stress level with fan operation in aerodynamic stall with a fitted stabilisation ring resulted in alternating stress increasing, and consequently, the mechanical safety factor reducing from 2.5 to 1.1. This increase is significant within the context of a smoothly rising fan characteristic that provided little indication that alternating stress had increased.

The smoothly rising fan characteristic resulted in Bard [33] claiming that "unstable performance due to stalling is completely eliminated." As Bard conducted a purely aerodynamic programme, making no measurement of steady or alternating stress, we may assume that when Bard referred to "unstable performance" he was referring to unstable aerodynamic performance. However, the claim may have resulted in some fan designers assuming that mechanical stress would also remain stable.

A tunnel ventilation fan with a nonstalling blade angle classically exhibits a continually rising characteristic. In effect, the fan blade aerodynamic loading is light enough that the fan does not suffer a classical aerodynamic stall. Alternating stress in a fan blade with a nonstalling blade stagger angle when operated in the unstable region of the fan characteristic does increase compared to the same fan operating in the stable region. Mechanical safety factor reduces from 2.4 to 1.5, and although any reduction in safety factor is undesirable, a safety factor of 1.5 is, nevertheless, high enough to make stalled operation possible, without suffering a fatigue-related mechanical failure.

The experimental results for both the stalling blade angle with fitted stabilisation ring and nonstalling blade angle with no stabilisation ring both result in mechanical safety factors that are less then the industry norm of 2.0. As such, we can regard both a stabilisation ring and nonstalling blade angle as methods to mechanically protect a tunnel ventilation fan in the event of an unforeseen stall event. If the fan application is one in which the fan will routinely drive into stall, then the prudent fan designer would increase the mechanical safety factor during stable operation to ensure that it did not fall below 2.0 during stalled operation.

The practice of selecting tunnel ventilation fans to accommodate pressure pulses within the stable part of the fan characteristic avoids the associated mechanical risk with operating tunnel ventilation fans in the stalled condition. However, this approach requires the tunnel ventilation system designer to foresee how the tunnel will operate for the life of the ventilation system. The current practice of fitting platform screen doors to historic metros has significantly increased the magnitude of pressure pulses, with the consequent risk that tunnel ventilation fans that the pressure pulse did not previously drive into stall will occur.

Additionally, fans that are correctly selected to operate within the stable part of their characteristic at 100% speed will likely drive into stall if operated at 50% design speed and certainly drive into stall if operated at 25% design speed. In this research paper, the authors were able to demonstrate that for a tunnel ventilation fan with a stalling blade angle, without a fitted stabilisation ring, the mechanical factor of safety during stable operation is 2.3. When they operated the same fan at half speed in an aerodynamically stalled condition, the mechanical factor of safety is 2.5. As the mechanical factor of safety at half speed during stalled operation is higher then the mechanical factor of safety at full speed during stable operation, the authors conclude that users can operate this tunnel ventilation fan at half speed in an aerodynamically stalled condition without risk of a fatigue-related mechanical failure. At

25% design speed, mechanical factors of safety in both stable and aerodynamically stalled operation are sufficiently high that there is no risk of a fatigue related mechanical failure. The authors conducted the current study on a single fan, and it is, therefore, not possible to generalise the findings to all fan types. Despite the limitations of the current study, the sevenfold increase in alternating stress (from 2.27 MPa to 16 MPa) that the authors observed in the tested fan with a stalling blade angle and without a fitted stabilisation ring is consistent with the conclusions of Rippl's [1] research. Despite the difficultly in generalising results of the reported research, it is possible to observe that if the fan is to operate reliably, the mechanical design must account for the increase in alternating stress when the fan stalls. Not doing so could result in the fan's fatigue related mechanical failure.

When accounting for the increase in alternating stress when a fan stalls, and when assessing if that increase is acceptable, the tunnel ventilation fan designer must make an assumption about the maximum defect size in the fan blades and hub. Therefore, responsible fan manufacturers 100% X-ray inspect all fan blades and hub to verify that they do not exceed the fan designer's assumptions regarding maximum defect size. Last, the authors conclude that the three fan selection strategies classically used by tunnel ventilation system designers each have specific advantages and disadvantages. The optimum fan selection strategy in a specific application will, therefore, depend on the impact of fan diameter on plant room cost and the relative importance of fan initial cost and fan-operating cost.

ACKNOWLEDGMENTS

The authors conducted this research in the context of contract FW-DMA09-11 between Fläkt Woods Ltd and the Dipartimento di Ingegneria Meccanica e Aerospaziale, "Sapienza" University of Rome.

REFERENCES

1. A. Rippl, Experimentelle untersuchungen zuminstationaren betriebsverhahen an der stabilitarsgrenze eines mehrstufigen transsonischen verdichters, Ph.D. dissertation, Ruhr-Universität Bochum, 1995.

2. S. K. Ivanov, "Axial blower," Patent No. US 3,189,260, 1965.

3. J. T. Gravdahl and O. Egeland, Compressor Surge and Rotating Stall: Modeling and Control, Springer, London, UK, 1999.

4. I. J. Day and N. A. Cumpsty, "The measurement and interpretation of flow within rotating stall cells in axial compressors," Journal of Mechanical Engineering Science, vol. 20, pp. 101–114, 1978.

5. E. M. Greitzer, "Review—axial compressor stall phenomena," Transactions of the ASME, Journal of Fluids Engineering, vol. 102, no. 2, pp. 134–151, 1980.

6. F. K. Moore, "A theory of rotating stall of multistage axial compressors: parts I–III," Transactions of the ASME, Journal of Engineering for Gas Turbines and Power, vol. 106, no. 2, pp. 313–336, 1984.

7. H. W. Emmons, C. E. Pearson, and H. P. Grant, "Compressor surge and stall propagation," Transactions of the ASME, vol. 77, pp. 455–469, 1955.

8. N. A. Cumpsty, "Part-circumference casing treatment and the effect on compressor stall," ASME Paper No. 89-GT, 1989.

9. S. Bianchi, A. Corsini, and A. G. Sheard, "Detection of stall regions in a low-speed axial fan. — part i: azimuthal acoustic measurements," in Proceedings of the 54th American Society of Mechanical Engineers Turbine and Aeroengine Congress, vol. 3, pp. 169–179, Glasgow, UK, June 2010.

10. A. G. Sheard, A. Corsini, and S. Bianchi, "Method of detecting stall in an axial fan," Patent No. GB 2 468 571 B, 2010.

11. A. G. Sheard, A. Corsini, and S. Bianchi, "Stall warning in a low-speed axial fan by visualization of sound signals," Journal

of Engineering for Gas Turbines and Power, vol. 133, no. 4, Article ID 041601, pp. 1–10, 2011.

12. C. C. Koch and L. H. Smith, "Loss sources and magnitudes in axial-flow compressors," Transactions of the ASME, Journal of Engineering and Power, vol. 98, no. 3, pp. 411–424, 1976. ·

13. H. Saathoff and U. Stark, "Tip clearance flow induced endwall boundary layer separation in a single-stage axial-flow low-speed compressor," ASME Paper No. 2000-GT-0501, 2000.

14. S. A. Khalid, A. S. Khalsa, I. A. Waitz et al., "Endwall blockage in axial compressors," Transactions of the ASME, Journal of Turbomachinery, vol. 121, no. 3, pp. 499–509, 1999.

15. H. Vo, C. S. Tan, and E. M. Greitzer, "Criteria for spike initiated rotating stall," ASME Paper GT 2005-68374, ASME Turbo Expo, Reno-Tahoe, Nevada, Nev, USA, 2005.

16. M. M. Bright, H. Qammar, H. Vhora, and M. Schaffer, "Rotating pip detection and stall warning in high-speed compressors using structure function," in Proceedings of AGARD RTO AVT Conference, Toulouse, France, May 1998.

17. T. R. Camp and I. J. Day, "A study of spike and modal stall phenomena in a low-speed axial compressor," Journal of Turbomachinery, vol. 120, no. 3, article 393, 9 pages, 1998.·

18. A. Deppe, H. Saathoff, and U. Stark, "Spike-type stall inception in axial flow compressors," inProceedings of the 6th Conference on Turbomachinery, Fluid Dynamics and Thermodynamics, Lille, France, 2005.

19. M. D. Hathaway, "Passive endwall treatments for enhancing stability," Report NASA/TM-2007-214409, 2007.

20. G. L. Wilde, "Improvements in or relating to gas turbines," Patent No. 701,576, 1950.

21. R. C. Turner, "Improvements in or relating to gas turbines," Patent No. 826,669, 1955.

22. R. G. Griffin and L. H. Smith Jr., "Experimental evaluation of outer case blowing or bleeding of a single stage axial flow compressor, part I—design of rotor blowing and bleeding configurations," NASA Report CR-54587, 1966.

23. E. E. Bailey and C. H. Voit, Some observations of effects of porous casings on operating range of a single axial-flow compressor rotor. Report NASA-TM-X-2120, 1970.

24. D. C. Prince, D. D. Wisler, and D. E. Hilvers, "Study of casing treatment stall margin improvement phenomena," NASA Report CR-134552, 1974.

25. D. C. Wisler and D. E. Hilvers, "Stator hub treatment study," NASA Report CR-134729, 1974.

26. H. Takata and Y. Tsukuda, "Stall margin improvement by casing treatment—its mechanism and effectiveness," Journal of Engineering for Power, vol. 99, no. 1, article 121, 13 pages, 1977. ·

27. E. M. Greitzer, J. P. Nikkanen, D. E. Haddad, R. S. Mazzawy, and H. D. Joslyn, "A fundamental criterion for the application of rotor casing treatment," Journal of Fluids Engineering, Transactions of the ASME, vol. 101, no. 2, pp. 237–243, 1979.

28. S. G. Koff, R. S. Mazzawy, J. P. Nikkanen, and A. Nolcheff, "Case treatment for compressor blades," Patent No. US 5,282,718, 1994.

29. S. J. Khalid, "Compressor endwall treatment," Patent No. US 5,520,508, 1996.

30. N. A. Nolcheff, "Flow aligned plenum endwall treatment for compressor blades," Patent No. US 5,586,859, 1996.

31. F. S. Gelmedov, E. A. Lokshtanov, L. E.-M. Olstain, and M. A. Sidorkin, "Anti-stall tip treatment means," Patent No. US 5,762,470, 1998.

32. S. Karlsson and T. Holmkvist, "Guide vane ring for a return flow passage in axial fans and a method of protecting it," Patent No. US 4,602,410, 1986.

33. H. Bard, "The stabilization of axial fan performance," in Proceedings of the Institution of Mechanical Engineers Conference C120/84 on the Installation Effects in Ducted Fan Systems (IMechE ‹84), pp. 100–106, 1984.

34. Y. Miyake and T. Inaba, "Improvement of axial flow fan characteristics by means of separators," Journal of Turbomachinery Society of Japan, vol. 13, pp. 746–752, 1985.

35. N. Yamaguchi, M. Ogata, and Y. Kato, "Improvement of stalling characteristics of an axial-flow fan by radial vaned air separators," Journal of Turbomachinery, vol. 132, no. 2, Article ID 021015, 10 pages, 2010.

36. C. S. Kang, A. B. McKenzie, and R. L. Elder, "Recessed casing treatment effects on fan performance and flow field," ASME Paper No. GT-95-197, 1995.

37. A. G. Sheard, A. Corsini, S. Minotti, and F. Sciulli, "The role of computational methods in the development of an aero-acoustic design methodology: application in a family of large industrial fans," inProceedings of the 14th International Conference on Modelling Fluid Flow Technologies, pp. 71–79, Budapest, Hungary, September 2009.

38. C. A. Wasserbauer, H. F. Weaver, and R. G. Senyitko, "NASA low-speed axial compressor for fundamental research," NASA Technical Memorandum 4635, 1995.

39. W. C. Young, Roark's Formulas for Stress and Strain, McGraw-Hill, New York, NY, USA, 1989.

40. W. Z. Gerber, "Calculation of the allowable stresses in iron structures," Bayer Architecture & Engineering, vol. 6, pp. 101–110, 1874.

Chapter 3

The Influence of Inlet Air Cooling and Afterburning on Gas Turbine Cogeneration Groups Performance

Ene Barbu[1], Valeriu Vilag[1], Jeni Popescu[1], Bogdan Gherman[1], Andreea Petcu[1], Romulus Petcu[1], Valentin Silivestru[2], Tudor Prisecaru[1], Mihaiella Cretu[1], and Daniel Olaru[1]

[1]INCDT COMOTI, Bucharest, Romania
[2]Politehnica University, Bucharest, Romania

INTRODUCTION

Usually, cogeneration is defined as combined production of power and thermal energy from the same fuel source, represented by natural gas, liquid fuel, refinery gas, etc. In conventional energy production the efficiency is approximately 40 %, but through cogeneration it can reach even 90 %. Fuel supply and increased performance requirements, environment concerns, continuously variable market conditions have contributed to the development of the gas turbines. The performances, exploitation costs, safety in operating conditions have made these installations to be selected for cogeneration processes.

STATE OF ART

Gas turbine systems operate on the ideal thermodynamic cycle (consisting in two isentropic and two isobars) represented by Brayton cycle. The real Brayton cycle consists in quasiadiabatic expansion and compression processes, but unisentropic, and the heat transfer processes are not isobar processes, due to flow pressure losses. In addition, the air and hot gases are not perfect gases and not have the same flow rates. Brayton cycle thermal efficiency depends on: compression ratio; ambient temperature; air temperature at turbine inlet; compressor efficiency and turbine components efficiency; blade cooling requirements; increased performance systems (exhaust gases heat recovery, intercooling, intake air cooling, afterburning implementation, fluids injection – water/steam, etc.). The main parameters that define the operating thermodynamic cycle of gas turbine installations (usually disclosed by the suppliers in catalogues) are the temperature at the gas turbine inlet (T_3) and the compression ratio. Generally, gas turbine manufacturers declare performances without taking into consideration the inlet and outlet pressure losses. Gas turbine installations performances are affected by the variation of these parameters as follows [2]: temperature increase at the gas turbine inlet leads to an increase in power and efficiency; the efficiency becomes maximum at a given value of the compression ratio (in T_3=const. hypothesis); there is a value of the compression ratio for which the power is maximum (T_3 and compressor intake air flow rate remains constant). The productivity of a gas turbine cogeneration group depends on the quantity of heat recovered from the turbine exhaust gases (approximately 60-70 % from the fuel energy). This is achieved by adding a heat recovery steam generator in order to supply hot water or steam. Determining factors in total efficiency of the cogeneration group are the gas turbine outlet temperature and the temperature at the stack of the heat recovery steam generator. The combination temperature at the gas turbine inlet – compression ratio determins the outlet temperature. The gas turbine, being located at the upstream of the heat recovery steam generator, significantly influences the cogeneration group performances. The air is induced by the gas turbine compressor in ambient conditions imposed by the location of the cogeneration plant. Compressor inlet temperature and intake air density dictates mechanical work required by the compression process, the fuel and quantity of fuel to be used in order to obtain the necessary temperature at the gas turbine inlet (T_3). Consequently, output power, efficiency, exhaust gases mass flow and outlet temperature (respectively the quantity of heat recovered) are influenced by ambient conditions [3]. The location of the gas turbine cogeneration plant imposes climatic conditions and requires adequate technical solutions in order to ensure performances. Generally, for cogenerative applications, the gas turbine is designed to operate in standard conditions, established by the International Standards Organization and defined as ISO conditions: 15 °C, 1.013 bar and 60 % humidity. During summer season air temperature rises and its density decreases, leading to a decrease in the intake air mass flow; consequently decreases and power output because it is proportional to the intake air mass flow rate. Without taking supplementary measures, both gas turbine output power and efficiency drop. In the scientific literature there are various papers that deal with the gas turbine's performance dependence of the intake air temperature variation [3-10]. In [4] it is shown that: an increase of 10 °C at the compressor inlet reduces the gas turbine outlet power with 18%; in comparison with the operation during winter season, the increase of ambient temperature leads to a decrease in gas turbine plants power output with 25-35%, also leading to an average increase of the consumption of 6%. The effect of intake air temperature over the performances differs from one gas turbine to another, but, generally, aeroderivative gas turbines are more sensitive to this phenomenon than the industrial gas turbines [5]. During summer season, when the days are long and hot, the power requirements increase for the residential spaces ventilation, offices, store rooms, etc. Additional energy consumption can be ensured by starting other backup groups, or compensating the loss of power through various other methods. The usual compensation methods of power loss are [6, 7]: compressor inlet air cooling (pre-cooling), intermediate cooling (intercooling), using recovery cycle. Mainly there are two basic compressor inlet air cooling methods: evaporative cooling (with evaporative media cooling or water injection in the inlet air-fogging); refrigeration system cooling [8]. For a 79 MW gas turbine, equipped with a fogging cooling system, the researches conducted at Mashhad (in Iran) showed that during a day, the maximum increase in power is achieved in the afternoon, when the temperature is higher and relative humidity is lower [9]. Inlet air cooling systems analysis in order to be applied to a gas turbine V94.2, in terms of efficiency increase, led to the conclusion that the fogging cooling system meets the design requirements and leads to an increase in power output of approximately 6 MW [10]. With the help of GT PRO software the performances of a 100 MW gas turbine model were analyzed, for a various types of inlet air cooling systems, and it had been reached that a decrease of air temperature of 1 ^{0}C (in the 25-35 ^{0}C interval) leads to an increase in power output of approximately 0.7 MW [11]. For a gas turbine cycle, with intermediate cooling (intercooler), the decrease of inlet air temperature causes the output power to rise and the intercooling leads to a 5-9% gain of power and a 8% reduction in fuel consumption [12]. Reduction of fuel consumption represents a priority both for industrial gas turbine manufacturers and also for the civil aviation. The search is on for new materials that meet the requirements imposed by the higher strains of the gas turbines [13] and also the development of new technologies, including technological transfer from aviation domain to power generation domain. Thus, in aviation, afterburning is used in order to increase traction of supersonic engines. The introduction of afterburning into cogenerative applications leads to an increase in flexibility and global efficiency of the cogeneration group. Afterburning application is possible due to the fact that exhaust gases at turbine outlet have a 11-16% (volumes) content of oxygen [14]. The afterburning installation, located between the gas turbine and the heat recovery steam generator, interacts with the gas turbine but influences the heat recovery steam generator operation especially [14, 15]. To increase the performance of gas turbine cogeneration groups research focused specifically on [16]: increase in burning temperature; increase in compression ratio; improving the methods of design, cooling and burning technologies, and also advanced materials; technological transfer from the aviation domain in the industrial gas turbine domain and conversion of aviation gas turbine (with outdated lifetime) to energy conversion; integrated systems (combined cycles, compressor inlet air cooling, intercooling, turbine exhaust gases heat recovery, afterburning implementation, chemical recovery, etc.). Following the direction displayed in the field, the chapter integrates data from scientific literature with research developed at INCDT COMOTI Bucharest, regarding gas turbine inlet air cooling and afterburning application, as base methods for increasing performances and flexibility of cogenerative group.

INFLUENCE FACTORS AND METHODS OF INCREASING PERFORMANCES IN GAS TURBINE COGENERATIVE GROUPS

For a combined cycle (considering as variables ambient temperature, gas turbine outlet temperature and stack temperature) it is shown that the dominant factor in global efficiency rise is stack temperature [17]. Obtaining a high efficiency involves the optimization of the entire cogenerative plant (gas turbine, afterburning installation, heat recovery steam generator, etc.). The efficiency must be maintained even at partial loads (even under 50%) in variable conditions modification. In general, although the target is obtaining a maximum efficiency, nevertheless an adequate flexibility to process requirements is desired, the afterburning installation contributing to this.

Influence Factors

Ambient parameters (humidity, pressure, temperature) can vary significantly depending on geographic location and season, affecting air density and implicitly the gas turbine cogenerative group performances. In the past, the effect of air humidity was neglected but the increase in gas turbine cogenerative groups power and the introduction of water/steam in the combustion chamber made this effect to be reconsidered. Thus, some authors [18] consider that air relative humidity (even at temperatures higher than 10 °C) has a neglectable influence over the gas turbine output power (as the other performance parameters). This leads to the fact that in some calculus (especially when the results are presented in correlation to ISO conditions) the variations in atmospheric humidity and pressure to be neglected. Others consider that due to the fact that water content modifies thermodynamic properties of inlet air (density, specific heat), at certain gas turbines (depending on specific processes) the performances may increase when humidity rises and in the case of some gas turbines the performances may decrease in the same conditions [19]. However, the increase in relative humidity leads to a significant reduction of NO_x emissions [20].

Ambient pressure is defined by the conditions from plant location, altitude modification leading to air density modification and implicitly to power output variation. Thus, 3-4% losses occur for each 304.8 m (1000 ft) rise in altitude [21].

Power and efficiency of the gas turbine group decrease along with ambient temperature, in figure 1 linear approximate variations being presented. Specific fuel consumption increases with the ambient temperature rise [22]. Gas turbines operate on a wide variety of gaseous fuels (natural gas, liquefied natural gas-LNG, liquid petroleum gas-LPG, refinery gas, etc.) and liquid fuels (kerosene. no. 2 diesel, jet A, etc.). Using a certain type of fuel for the gas turbine has a profound impact both on the design and also on material selection. Usage of liquid fuels imposes: ensuring burning without incandescent particles and residues on the combustor and turbine; reducing hot gas corrosive effect due to aggressive compounds (sulphur, led, sodium, vanadium, etc.); resolving pumping and pulverization (filtering, heating, etc.) issues. In case of using gaseous fuels, a simpler solution is presented due to their higher thermal stability, higher heating power, lack of ash and smut. However, in order to ensure pressure level (required by the gas turbine, afterburning installation, etc.), water and various impurities elimination implies a control-measuring station for the gaseous fuels used (natural gas in the case of cogenerative plant 2xST18 – Figure 2). Although the main fuel for the operation of gas turbine cogenerative groups is natural gas, the economic rise and environmental requirements issued an alternative. The gas turbine can be designed to operate on a variety of fuels, but the rapid transition to other fuel operation, without machine damage or exceeding the level of emissions, still remains an issue subjected to study.

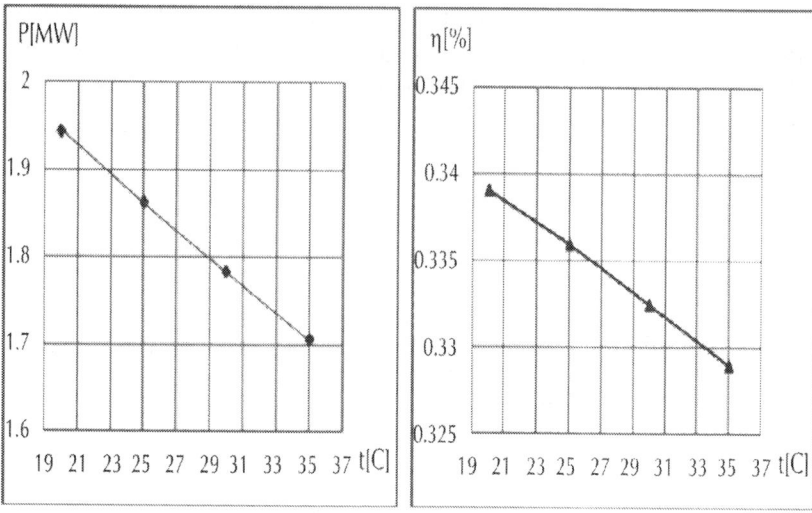

Figure 1: Power output (left) and efficiency (right) variations versus ambient temperature, at a simple gas turbine cycle.

Interchangeability at gaseous fuel gas turbine cogenerative groups, represents the ability to change a fuel with another one, without affecting the application or the equipment in which the gaseous fuel is burned [15]. At a constant fuel composition (in the case that the holes through which the fuel passes to the burner have fix dimensions), the quantity of heat delivered by the burner is proportional with the mass flow and heating power. When the composition varies and it is the problem of replacing a fuel with another equivalent, Wobbe index (after John Wobbe-engineer and mathematician) is used as comparison criteria. It is defined as ratio of the lower heating value (LHV) and the square root of relative fuel density respectively, in relation to air (d_{rel}):

$$Wo = LHV / (d_{rel})^{0.5} \tag{1}$$

$$d_{rel} = \rho_{comb} / \rho_{air} \tag{2}$$

Thus, two gaseous fuels (with different chemical compositions), with the same Wobbe index, are interchangeable and the quantity of heat delivered to the equipment is equivalent at the same fuel supply pressure. In order to take into account the fuel temperature, a Wobbe temperature corrected index can be used. According to [23], two gaseous fuels are interchageable if the following relation is satisfied:

$$\frac{\Delta p_2}{\Delta p_1} = \left(\frac{Wo_2}{Wo_1}\right)^2 \left(\frac{A_1}{A_2}\right)^2$$

(3)

where Δp_1 and Δp_2 represent the overpressures of gas 1 and gas 2 respectively, Wo_1 and Wo_2 – Wobbe indices of fuels 1 and 2, A_1 and A_2 – gaseous fuel injection nozzle surface areas.

Figure 2: Cogenerative plant 2xST 18 – Suplacu de Barcau (left) and afterburning installation (right) [14, 15].

Thus, the validation criteria of adequate replacement of one fuel with another equivalent fuel are given by: self-ignition, flame temperature (with a high influence on NO_x emissions forming), flame speed, flashback, efficiency, NO_x and CO emissions, flue gas dew point, etc. Resolving gas turbine cogenerative group fuels interchangeability, by developing high performance alternative fuel burning technologies, especially hydrogen, will have a major impact over system and environment efficiency. Thus, the studies conducted on more fuels (H_2, CH_4, C_3H_6, C_6H_6, CH_3OH) revealed that [24]: hydrogen and methyl alcohol have the same higher maximum efficiency than other fuels in the same operating conditions; hydrogen fuel has the lowest specific fuel consumption in comparison with other fuels, followed by methane, propen, benzene, and finally methyl alcohol; in the reheating cycle the increase in thermal efficiency is lower than the increased in the intercooling cycle; hydrogen fuel is ideal promising fuel in the gaseous plant which has greater thermal efficiency and greater improvement in the performance of modified gas turbine power plant occurred with intercooling and heat exchanger rather than simple and reheat cycle.

From the point of view of reusing aviation gas turbines for industrial purposes, the possibilities of using liquid fuels are limited, leading to the development of new technologies on gaseous fuels. Thus, the transition of a TV2-117A gas turbine engine from liquid to a gaseous fuel, in order to benefit from landfill gas energy value, has been conducted in many stages: transition from liquid fuel (kerosene) operation to gaseous fuel (natural gas) operation, thus obtaining TA2 gas turbine engine; the transition of TA2 gas turbine engine from natural gas operation to landfill gas operation, thus obtaining TA2 bio gas turbine engine. In order to obtain the two gas turbine engines (TA2 and TA2 bio), numerical simulations were conducted in CFD environment, constructive modifications and gas turbine test bench experiments in order to validate adopted solutions [15]. In this way, TA2 gas turbine engine was integrated in the structure of afterburning installation test bench facilities, from INCDT COMOTI Bucharest (figure 3).

Figure 3: TA2 gas turbine engine mounted on the stand (left) and afterburning installation (right).

Efficiency and reliability are two major parameters that are taken into account since the beginning of a new gas turbine cogenerative group design. In order to obtain a higher efficiency (in balance with cost and reliability), the design team must reach a balance between burning temperature rise and compression ratio, special material selection and complicated cooling systems, customer's specification, etc. Based on a continuum dialogue with Siemens customers, as far back as the predesigned stage, a 36 MWe SGT-750 gas turbine was elaborated [25]. It can be used both in cogenerative applications as well as driving different equipment, stable operation at partial load but also allows the transition to another fuel (operates on dual fuel). Since 2002, Siemens began focusing attention to reliability, so that they give up the high pressure tambour from the recovery boiler (usually used in order to prevent high thermal tensions, a long period of time is imposed in order to reach a certain temperature). Regarding efficiency, flexibility and emission reduction at gas turbine cogenerative groups, important steps were achieved towards: integrating low NO_x burners, lifecycle was analyzed in order to increase efficiency, maintenance interval was enlarged and the transition from one fuel to another was improved at polifuel groups.

Methods of Increasing Performances

The development of gas turbine application in various regions of the globe, encouraged researchers to find new methods of increasing performances and to apply new cooling technologies to compressor inlet air, specific to the location. Actually the advantages of inlet air technology application are represented by power losses prevention, losses that occur when ambient temperature exceeds 15 °C (ISO design temperature), and fuel usage efficiency. In general, the studies of selecting a new method of cooling air take into account each method (see chapter 1) and compare them with a reference case [8]. Evaporative cooling systems imply lower investments, operating and maintenance costs than refrigeration systems but the increase in gas turbine performances is lower. Evaporative cooling system is based on the transition of an air flow through a water soaked environment, system efficiency depending on the surface area of the water soaked environment exposed to air and exposure time. The system is efficient in low humidity regions. In the case of fogging system, the demineralized water is sprayed in the gas turbine inlet air flow, through high pressure nozzles (70 – 200 bar). Figure 4 shows a fogging system, located downstream of the filtering system, in a gas turbine. The air induced through the filtering system reaches the suction chamber, where the water, sprayed in fine droplets (<30 microns), is evaporated in the air flow and produces a cooling effect. In general, the water injected mass flow required represents 2 % from the inlet air, depending on the ambient parameters (temperature and humidity).

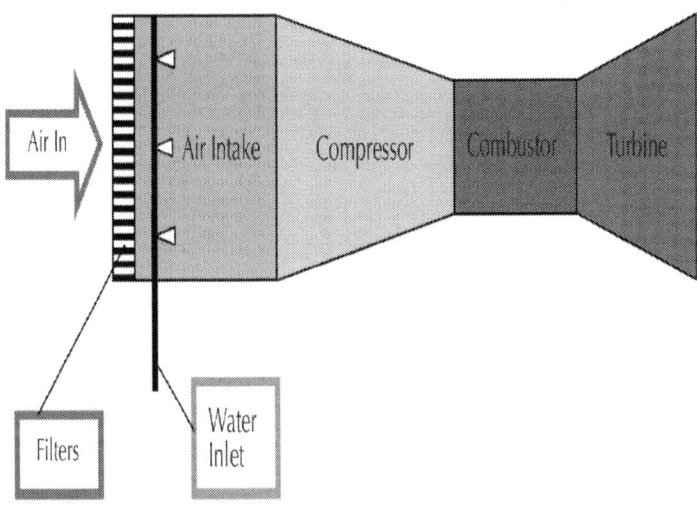

Figure 4: Fogging system scheme, located downstream of the filtering system.

Water injection system (with fog) has the advantage of lower pressure drops, higher efficiency, lower refurbishment costs but requires demineralized water, parasite loads occur due to water injection pumps and the control system is complex. Fogging systems, increasingly popular in time, are capable of cooling gas turbine intake air to the temperature of the wet-bulb thermometer, being more efficient than the ones with evaporative media. Although significant investments are required, refrigeration cooling systems (mechanical, absorption, storage) can practically maintain any temperature of gas turbine intake air, recommended in variable humidity regions. In general, fogging system are designed to operate in such a way that the formed droplets should evaporate until entering the compressor, in order to avoid blade damage. In some cases overspray phenomenon can occur (a certain amount of water would not evaporate until entering the gas turbine). In this way overspray phenomenon is similar with water injection between compressor stages (intercooling phenomenon occurring). In certain operating regimes it is possible that evaporative cooling (upstream of the compressor) combined with overspray phenomenon (intercooling) lead to compressor aerodynamic instability (although intercooling effect is added to evaporative cooling) [26]. Combining overspray phenomenon with the usage of a low power heating fuel, steam injection in the combustion chamber or a high level of blade wear can lead to catastrophic results for the gas turbine. Some authors [27] consider that at gas turbine cogenerative cycles, intercooling is recommended, in order to reduce high pressure compressor stage power consumption.

Thus, the factors that need to be taken into account when adopting a new technology of cooling the inlet air are [28]: plant configuration (gas turbine engine can be used within an open cycle, a cogenerative cycle, a combined cycle with high operation – intermittent or basic); the amount of compressed air per kW (requiring a small amount of compressed air the cooled air required decreases and, implicitly, exploitation costs decrease), the ambient air enthalpy at given conditions by the design (selecting a high enthalpy of air per kg of dry air leads to a large cooling system, with high exploitation costs); air cooling temperature (for each gas turbine engine model exists an optimum cooling temperatures at given environment conditions).

Although the integration of an afterburning installation (figure 2 and 3) can lead to an increase in cost (with approximately 10-15 % from the cost of the heat recovery steam generator) certain advantages are highlighted: an increase in steam amount at the heat recovery steam generator; thermal energy can be managed more easily; steam amount efficiency, depending on technological process requirements; the heat recovery steam generator can operate even at gas turbine shutdown; it can compensate ambient parameter variations; in some cases can contribute to NO_x emissions reduction by interacting with gas turbine emissions; can burn other fuels usually inadequate to gas turbines (biogas, furnace gas, flammable gases resulted from gasification, etc.).

THEORETICAL RESEARCH DEVELOPED AT INCDT COMOTI BUCHAREST

At INCDT COMOTI Bucharest, theoretical research approached issues concerning: processes regarding water spray in gas turbine intake; intercooling; influence of air cooling over gas turbine performances; thermogasodynamic processes from the afterburning installation's burner.

Thus, in order to understand the phenomenon of cooling air at water spraying through an impact body nozzle [29], numerical simulations in CFD environment were conducted, with the working domain consisting in air (gas) and water (liquid). In the first version the impact body (a cone with 1.2 mm base and generator lines) was not taken into account. In the second version the cone was positioned, firstly, at 0.4 mm towards the pipe's end, then moved at 0.8 mm, and 1 mm respectively. Sprayed water reaches in the calculus domain (a cylinder of 1m in diameter and 2.5 m in length) through a 1.2 mm diameter and 12 mm length pipe positioned along the cylinder's symmetry axis. In order to capture as accurately as possible the turbulence phenomenon that occurs in the cone and pipe area, at the same time with water flow sprayed around the cone, the mesh was refined (figure 5).

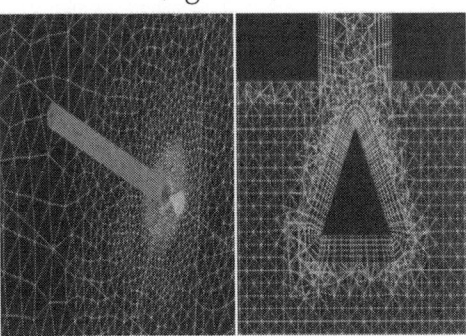

Figure 5: Calculus mesh at the inlet zone of sprayed water in the domain (left) and around the cone (right).

Also, along the pipe and cone walls boundary layers were created, in order for an accurate capturing of the flow near the walls. Water is sprayed into the atmosphere with an axial speed of 15 m/s, as droplets of 100 μm diameter, having a temperature of 288 K. Boundary conditions for the cone and pipe are of smooth adiabatic wall type. The turbulence model used was k-ε. Work methodology was based on the model described in [30]. Numerical simulation results are presented in figures 6 and 7, highlighting the spray without impact body and with impact body positioned at 1 mm distance from the pipe outlet.

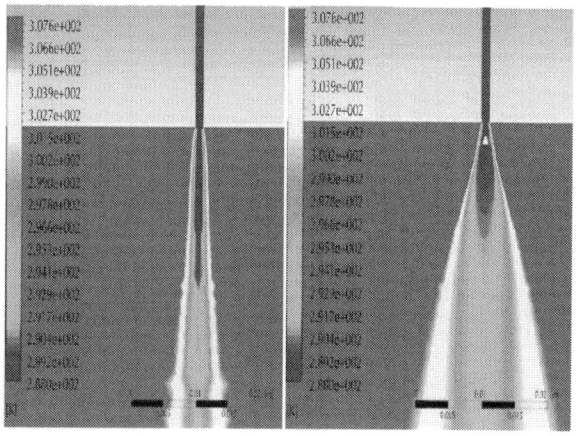

Figure 6: Detail regarding temperature field at water output: without cone (left) and with the cone positioned at 1 mm distance (right).

Figure 6 reveals that cone insertion leads to pronounced water jet flaring. Cone position modification has a minimum effect over the water spray flare angle. For a better observation of sprayed water jet along the calculus domain, 300 mm in diameter planes were created (comparable with TA2 gas turbine engine intake-figure 3) along the symmetry axis corresponding to z axis coordinate of 0.5 m, 1 m, 1.5 m, 2 m, and 2.5 m respectively.

Figure 7 shows that in case of numerical simulations without the cone, the water jet is directed along the symmetry axis, while in case of impact body numerical simulations the water jet is more dispersed, as expected. In the case of numerical simulations without the cone, the average temperatures decrease more rapidly.

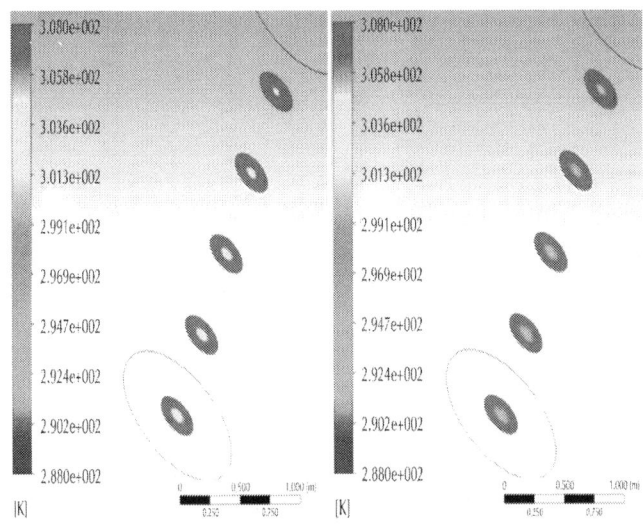

Figure 7: Temperature field in different cross planes: without cone (left) and with the cone positioned at 1 mm distance (right).

Compressor performances are influenced by flow steadiness at the rotor outlet, more specifically if the flow angle varies from hub to shroud than the flow within the diffuser will be unsteady and will lead to separations on the blade vane. Also, the flow at the tip of the blade and at the centrifugal rotor outlet is intensily distorted and unsteady. In order to quantify the impact of intercooling on a two stage compressor, a CFD study was conducted in which the second stage inlet temperature was reduced. Performance impact was monitored, in comparison with the case in which between the two compression stages no cooling was applied. Thus, two CFD analyses were conducted, in which the total inlet second rotor temperature was considered 460 K, for the case without cooling between the compression stages and 313 K, for the case with cooling.. The calculus mesh for this case is a structured mesh and has approximately 1 million elements. The walls are considered to be adiabatic, waterproof and velocity at wall is considered to be zero. In order to improve calculus time, a single channel has been considered (rotor consisting in 15 blades and 15 splitters), using the periodicity function. For spatial discretization was used a second order scheme. A compressible flow was considered in calculus, the governing equations being written as Reynolds Averaged, time and mass averaged. Shear Stress Transport k-ω was considered as turbulence model. SST k-ω model is based on tangential tensions transport. With the help of this turbulence model, accurate results and separation zone dimension (that forms under the influence of high pressure gradients) can be obtained. The results are presented in figures 8-10.

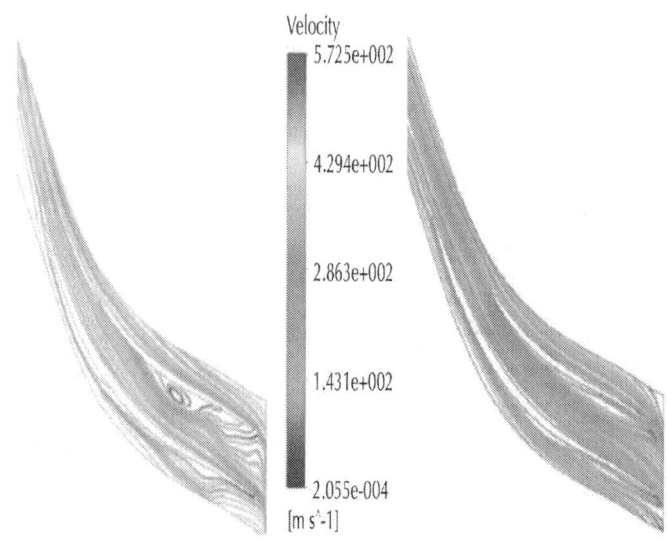

Figure 8: Streamlines, at 50 % of blade height, measured from the hub, in the case of cooling (left) and in the case without cooling (right).

Analyzing the CFD simulations results, it can be observed that along with temperature reduction at rotor inlet, streamlines in the rotor indicate that a recirculation zone is forming on the splitter. This shows that flow angles in the rotor have changed, fact that led to boundary layer separation in the case of intercooling (figure 8). Also, it can be observed (in meridional plane) the fact that Mach number rises at the rotor outlet, from 0.88 in the case without intercooling to 1.055 in the case with intercooling (figure 9).

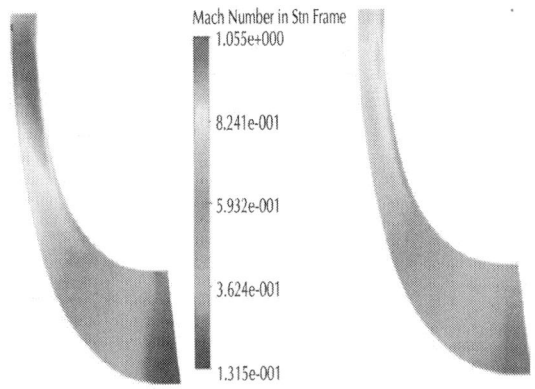

Figure 9: Mach number in meridional plane, in the cooling case (left) and in the case without cooling (right).

Another difference can be observed also in outlet rotor total pressure, the pressure increases in the case of intercooling (figure 10). Thus, at centrifugal rotor outlet, in the case of intercooling, an increase of 0.2 in Mach number can be observed, reaching 0.938. This indicates a transonic regime at rotor outlet. Corroborated with the fact that the flow angle at rotor outlet reached 12.72^0 from 17.713^0, it means that the existing stator must be redesign.

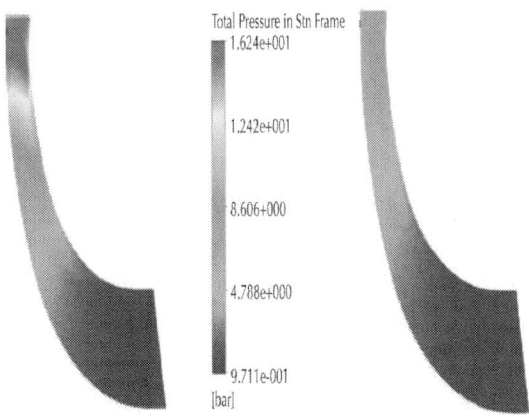

Figure 10: Total pressure in meridional plane, in the cooling case (left) and in the case without cooling (right).

In general, the performances of a turboshaft engine is compared with the ones given by ideal cycle calculus, thermodynamic analysis including thus, the ideal gas turbine work range. Gas temperature at turbine inlet, as gas turbine outlet section and outlet pressure, are equal in the case of ideal and real gas turbine [31].

Thus, the main configuration to be taken into account for thermodynamic analysis is the one for a monorotor turboshaft engine, with two high and low pressure compressor stages, coupled with a turbine that includes both gas generator stages, as well as power shaft supply stages. For performance calculus, the above configuration is considered (figure 11-left) and intercooling configuration (between compressor low and high pressure stages – figure 11 right). The heat exchanger, that provides intercooling, decreases air temperature in the high pressure compressor inlet section at 40 °C (313 K). Intercooling application decreases the mechanical work consumed by the compressor, without affecting the mechanical work produced by the turbine, leading to an increase in the mechanical load available at the output shaft [32]. In thermodynamic analysis, for the two gas turbine engine configurations presented above, a series of parameters were imposed for operational purposes. The calculus method used is in accordance with [33].

Common conditions to both configurations refer to: ambient temperature, whose variation influences gas turbine performances (shaft power, thermal efficiency, specific fuel consumption); gas temperature in the turbine inlet section, having the same value in all the cases presented, does not vary with ambient temperature, parameter imposed by turbine alloy properties; turbine outlet pressure, in order to ensure exhaust, in the context of pressure losses on the exhaust; burned gases pressure at the exhaust outlet so that the gases can pass through the heat recovery steam generator; coefficients of pressure, speed, energy losses with compressor and turbine efficiencies, and compression ratio (in initial configuration), low pressure compressor (in intercooling configuration) respectively are considered constant.

In addition, for the configration that includes intercooling, the following parameters are imposed: temperature at high compressor inlet, the same for all calculus cases, that allows the use of a low temperature resistant material and allows to obtain a high compression ratio at the same mechanical work consumed on the second compressor; the pressure loss coefficient in the heat exchanger; air pressure at combustion chamber inlet, that derives from total compression ratio imposition. In these conditions, the gas turbine output power variation depending on the inlet air temperature (for the various compression ratio), for simple configuration and intercooling and for different compression ratios, is presented in figure 12.

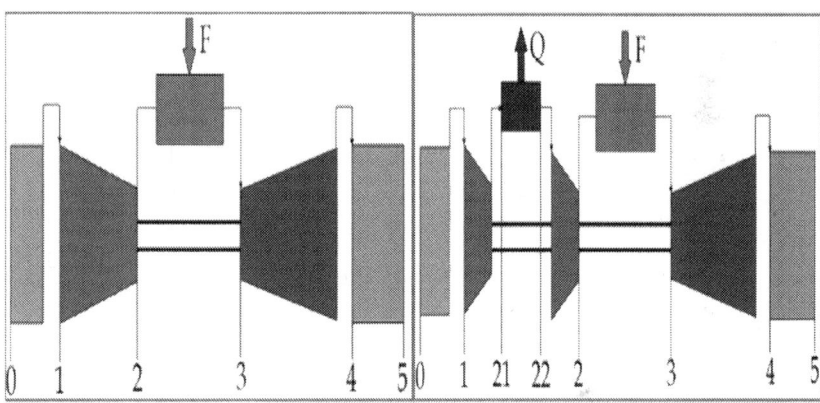

Figure 11: Gas turbine engine configurations – general schemes. (0 – intake inlet; 1 – compressor inlet (low pressure compressor); 21 – heat exchanger inlet; 22 – high pressure compressor inlet; 2 – combustion chamber inlet; 3 – turbine inlet; 4 – exhaust inlet; 5 – exhaust outlet).

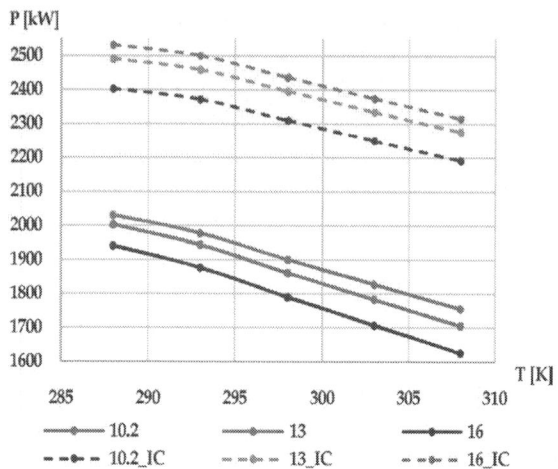

Figure 12: Gas turbine output power variation versus inlet air temperature (at the various compression ratio), for simple configuration and intercooling-IC.

It is noticed a decrease of shaft power and thermal efficiency, simultaneous with an increase in specific fuel consumption, at an ambient temperature increase, phenomenon that occurs for all cases in which thermodynamic analysis was conducted, both in the initial gas turbine configuration, as well as in intercooling. For the initial configuration, a pronounced variation occurs with ambient temperature increase, easily noticeable in the case of high compression ratios. In the case of intercooling gas turbine, performance variation is approximately linear for all the three compression ratios, with a decrease in power of 2.5 % for each 5 degrees of the environment.

In the field of afterburning, the theoretical research has soughed: to obtain a reduced pollutants afterburning installation, in comparison with the existing cogeneration power plant 2xST 18 (figure 3); afterburning installation flexibility when supplied with gaseous fuels; to study the influence of water spraying into the gas turbine's combustion chamber upon the afterburning [14, 15]. Thus, the numerical simulations performed in CFD environment have showed that (at nominal conditions), by modifying the afterburning module of the cogenerative plant 2xST 18, the NO_x concentration in the exhaust gases is lowered three times [14, 15]. The base module of the cogenerative plant has been mainly modifying by flaring at 15° and by introducing a concentrator.

INLET AIR COOLING INSTALLATION AND AFTERBURNING INTEGRATION WITHIN GAS TURBINE COGENERATIVE GROUP — FUTURE RESEARCH

Gas turbines usually function with high air excess, between 3 and 6 at nominal regime, and even higher at partial-load regimes [34]. This enables the use of an afterburning module, but the control of the air/fuel mixing requires the minimization of the air excess. Thus, reducing the exhaust emissions must be balanced against providing the cogenerative group's performances in a flexible manner.

Efficiency-emissions Binom

The main pollutant emissions produced by gas turbines consist of: nitrogen oxides (NO_x), carbon monoxide (CO), volatile organic compounds (VOCs). Gas turbines typically operate at high loads which make their design for optimum combustion and maximum efficiency to be made at nominal load. Controlling the concentration of all pollutants is difficult in the conditions of variable loads operating. In high loads operating regimes the concentration of NO_x is higher, while in lower loads operating regimes (under 50 %) the thermic efficiency decreases and the concentrations of CO and volatile organic compounds increase. Thus the factors which determine the formation of pollutants in the exhaust gases are [14, 15]: the temperature and the excess of air in the primary zone; the process homogenization degree in the primary zone; the combustion process products residence time; the "freezing" characteristic of the reaction near the fire tube; etc. To decrease the emissions of NO_x is necessary to reduce the temperature in the area in which the combustion reaction takes place and in the high temperature zones, respectively to rethink the distribution of the air flows (combustion in stages). The final version of the gas turbine's combustion chamber will be a compromise between the level of the pollutant emissions, performance and flexibility. Depending on the combustion temperature, in figures 13 and 14 are represented the NO_x and CO emissions levels typical for a class of industrial gas turbines, using different fuels and at various operating regimes [35]. The high CO emissions level indicates an incomplete combustion and a decrease in efficiency. From figures 13 and14 it can be observed that for temperatures up to 760 °C, the levels of NO_x and CO are comparable (especially for natural gases), while in the case of higher temperatures the NO_x emissions increases rapidly, while the CO emissions level remains practicly constant. By comparison, in the case of micro gas turbines, operating at 70-100 % loads, the CO emissions are low (figure 15) but they increase fast when operating at under 70 % load [36]. In the case of micro gas turbines the NO_x emissions level are low over a wide range of operating regimes (30-100 % load).

Adding a heat recovery steam generator and a cooling system for the intake air (with fog) could be a solution for increasing the performance and decreasing the influence of high temperature in the summer. The high content of vapours in the combustion gases (by injection of water/steam) leads to: acid corrosion (when using fuels which contain sulfur); increase thermal solicitations of the combustion chamber, reduce the heat recovery level, etc. The exhaust gases flow at the gas turbine exit is turbulent and uneven in the transversal section. Thus there might appear reverse flows in some areas of the heat recovery steam generator transveral section. The unevenness of the flow at the combustion chamber exit and the variation of the exhaust gases compozition influence the functioning of the afterburning module. Thus the afterburning is influenced in terms of efficiency, pollutants, flame stability but also in terms of corossion of the elements subject to the action of the exhaust gases.

Generally, for a good design of the exhaust gases flow into the heat recovery steam generator, the following factors must be taken into acount [15]: the geometry of the gas turbine exhaust and its direction; the size of the heat exchange surfaces; the position of the afterburning; masic flow and mean speed at the gas turbine exit; local speeds near the walls and at the entrance of the first heat exchange surface. In general the gas turbine's evacuation is not directly coupled with the heat recovery steam generator; after leaving the evacuation of the gas turbine (in the case of the cogenerative power plant 2xST 18 – Suplacu de Barcau, figure 2), the exhaust gases pass through a silencer, a by-pass assembly, an adaption section to the afterburning and then they reach the afterburning chamber. An uniform distribution of the flow in the transversal section insures a proper functioning of the heat recovery steam generator, especcially of its overheater. This creates the necessary premises to ensure low emissions for the cogenerative group. If the exhaust gases coming from the turbine or the air flow are not evenly distributed, in the same way as the fuel, serious variations of the temperature can appear downstream of the burner.

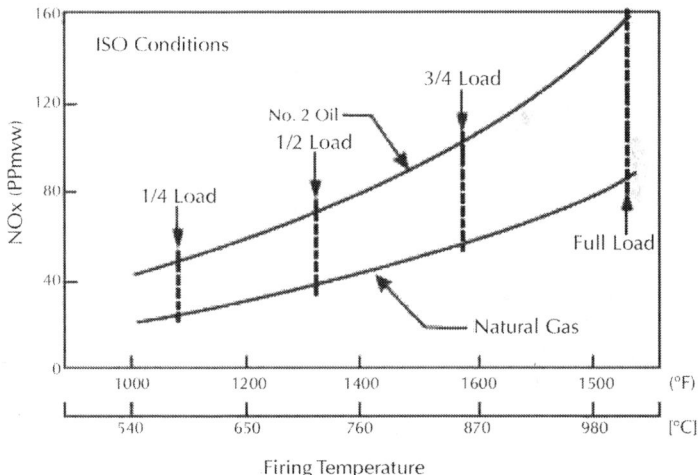

Figure 13: NO_x emissions variation for a class of industrial gas turbines - © 2010 Richard ‹Layi Fagbenle. Originally published in [35] under CC BY-NC-SA 3.0 license.

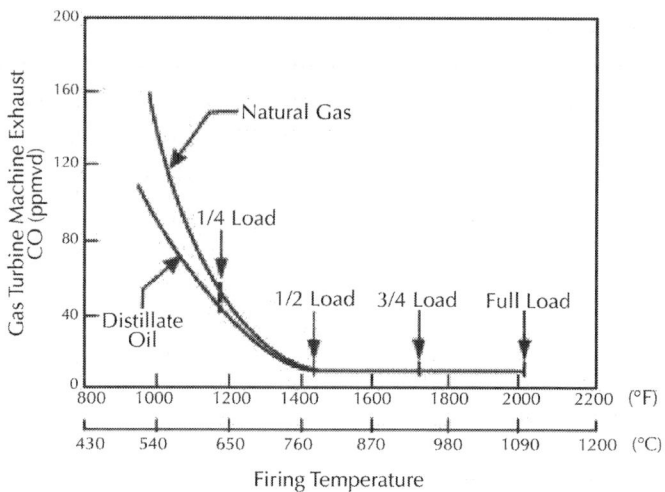

Figure 14: CO emissions variation for a class of industrial gas turbines - © 2010 Richard ‹Layi Fagbenle. Originally published in [35] under CC BY-NC-SA 3.0 license.

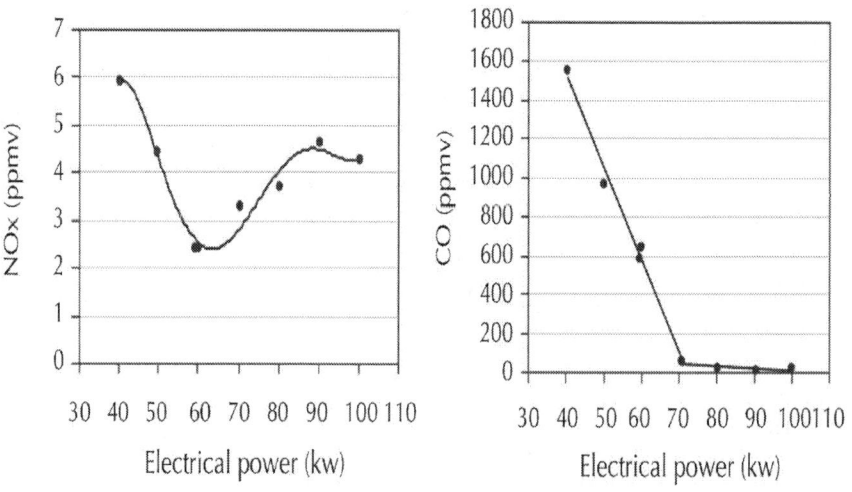

Figure 15: NO_x (left) and CO (right) emissions variation for a class of micro turbines - © 2010 Flavio Caresana, Gabriele Comodi, Leonardo Pelagalli, Sandro Vagni. Originally published in [36] under CC BY-NC-SA 3.0 license.

In general, the variation of the speed (upstream of the burner), on 90 % of the burner section, mustn't exceed ± 15 % of the mean speed on the whole transversal section. In reality the exhaust gases temperature, downstream of the burner, will never be perfectly uniform. Even with a perfect distribution of the gas flow in the turbine, upstream of the burner, the temperature of the gases in the zone of each afterburning module will be higher than the temperature in the areas between the modules. These requirements must fall in the market trends, where (in terms of fluctuation of the electricity and fuel prices) the flexibility in functioning has become a major subject, new concepts being imposed. Increased flexibility aims [37]: rapid start and stop; rapid change of the load; increase reliability in the case of quick start and load predictability; frequency control and of auxiliary services.

Experimental Research Conducted at INCDT COMOTI Bucharest

The experimental research carried out at INCDT COMOTI Bucharest, in close relation with the theoretical research (see chapter 4), have concentrated on: obtaining an afterburning installation on gaseous fuel, with low emissions; the realization of water spraying systems and their testing; experimenting water injection in the intake device of the gas turbine, at nominal functioning regime.

The numerical simulations carried out in CFD environment show that, by modifying the afterburning module of the cogenerative power plant 2xST 18, three times lower NO_x emissions can be obtained at nominal functioning regime [14, 15]. Up until now there have been carried out comparative experiments (figure 16) using gaseous fuel, at partial load (3 % of the nominal load). Thus the experiments at partial load show a reduction of the NO_x emissions with 30 %. From figure 16, right, it can be observed that the flame better fills the fire tube of the new afterburning module, and the temperature field is more uniform, being in good correlation with the results.

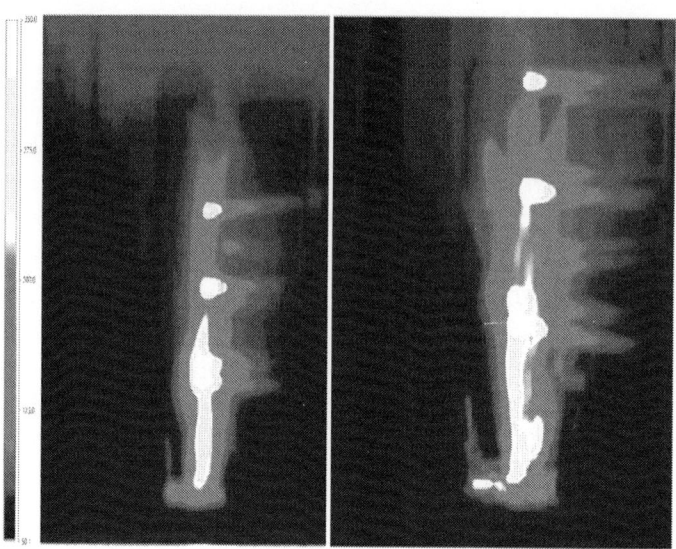

Figure 16: Comparative analysis, in infrared, of the afterburning modules (the existing module on the 2xST power plant – left; numerical simulation result – right) [15, 38].

Before beginning the experiments regarding the effects of water injection in the intake device of the TA2 gas turbine, there has been studied the form of the water spray jet using a liquid fuel atomizer from TV2-117A gas turbine (figure 17). In these tests there has been chosen the atomizer circuit which ensures a droplet diameter (figure 17 left) comparable with the dimension of the gas turbine's intake device (figure 18 left). During the tests the water (untreated) has been sprayed at a pressure of 30±0.5 bar, the atomizer being position on the TA2 gas turbine axis. The distance between the atomizer and the gas turbine intake has been varied between 1500 mm and 2000 mm. The testing rig is composed by: TA2 gas turbine, natural gas afterburning installation – positioned on the vertical, water spraying instalation, command and data acquisition chamber. The emissions measurement has been realized using the Horiba PG 250 gas analyzer, positioned on the stack (the exit of the afterburning chamber –figure 18 right). The experiments conducted on the TA2 gas turbine, using gaseous fuel, have been carried out at starting regime (10500±6 rpm; 75±5 Nm³/h natural gas flow).

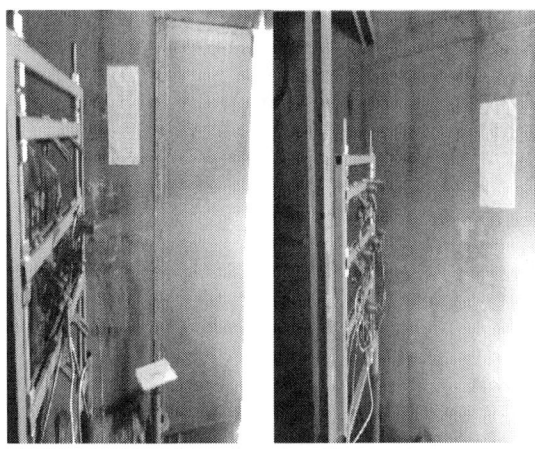

Figure 17: Water spraying tests, using a liquid fuel atomizer from TV2-117A gas turbine.

The mean temperature T_{3M} (the gases mean temperature before the gas turbine) has been determined using 17 thermocouples (figure 19 left), using the methodology presented in [15]. The free turbine speed has rezulted from the vibration analysis, data acquisition being realized unsing module IEPE BNC Ni9233 (National Instruments). The exhaust gases temperature (T_5 – figure 19 right) has been measured using a thermocouple.

Figure 18: Experiments regarding the injection of water in TA2 gas turbine (left) and gas analyser PG 250 (right).

The experiments included several series of 20 minutes in which the gas turbine functioned at starting regime (without the afterburning module), using natural gases as fuel (for 10 minutes the gas turbine functioned without water injection, followed by 10 minutes of functioning with water injection). The results regarding NO_x emissions during the experiments conducted on the TA2 gas turbine using natural gases as fuel, with/without water injection (the atomizer is positioned on the gas turbine's axis, at 1500 mm from the gas turbine's intake device) are presented in figure 20. The results obtained so far regarding water injection in the intake device of TA2 gas turbine (for power increase) are more qualitative, although by introducing water the speed of the free turbine is increased with about 20 rpm. Simultaneously with the introduction of water, temperatures T_{3M} and T_5 decrease with about 5 °C, and the NO_x emissions decrease with about 4 ppm. It is expected that the influences will be more conclusive when the load of the gas turbine and the water quantity are increased.

Figure 19: The thermocouples position for determining T_{3M} (left) and T_5 (exhaust device exit - right).

After about 10 series of experiments, of 20 minutes each, the boroscopic examination of the gas turbine didn't revile any significant deposits. Possibly in the future it will be necessary to appropriately treat the water injected in the intake device.

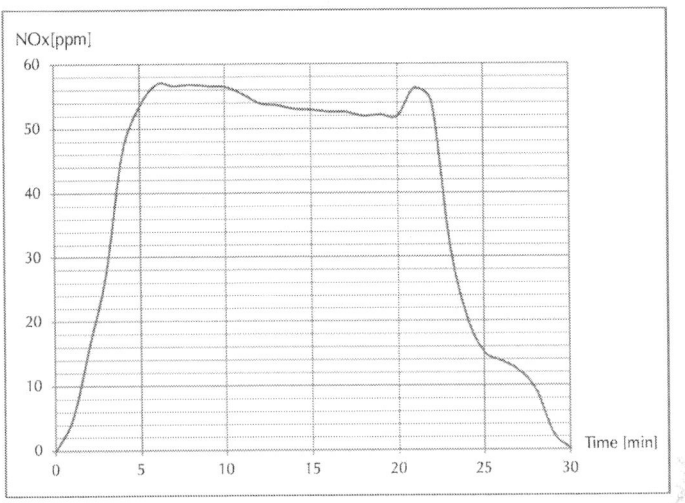

Figure 20: NO_x emissions variation during the experiments carried out on TA2 gas turbine, with water injection in the intake device.

Future Research

Future research to be conducted at INCDT COMOTI Bucharest will follow the general context given by the efficiency-emissions reduction-flexibility triad, by numerical simulations and experiments concerning: processes that take place during the cooling of the intake air; combustion in the gas turbine and afterburning installation; increase of the efficiency and pollutants reduction; flexibility regarding the used fuels.

CONCLUSIONS

- The selection of a location for a gas turbine cogenerative plant imposes climatic conditions and demands adequate technical solutions to meet performance requirements, especially during summer season when inlet air temperature rises, leading to a decrease in power output and efficiency.

- Water content modifies thermodynamic properties of intake air (density, specific heat) affecting power output and heat mass flow resulted from the gas turbine. If in the past air humidity was neglected, in present day cogenerative gas turbine power increase, but also water/steam injection impose the need for it to be taken into account.

- The main research directions in the area of cogenerative groups with gas turbines efficiency increase are: combustion temperature increase; compression ratio increase; improvement of design methods, combustion technology and advanced materials; technological transfer for aviation domain to industrial turbines domain; integrated systems (combined cycles, intake air cooling, exhaust turbine gases heat recovery, afterburning, etc.).

- Determinant factors concerning the overall efficiency of the cogenerative group are: gas turbine exit temperature, temperature at the heat recovery steam generator stack; ambient environment temperature. For these the most influential factor upon the increase of the overall efficiency is the temperature at the heat recovery steam generator stack.

- Operating flexibility of equipment has become a major subject. Gas turbines are designed to function generally at nominal regime, in maximum efficiency conditions and minimum pollutants. At cogenerative groups with heat recovery steam generator, for producing technological steam, is preferable that the flexibility to the process demands to be achieved using afterburning installation.

- Theoretical and experimental research conducted at INCDT COMOTI Bucharest, allowed: to be showed that, in the case of a gas turbine with intercooling, the performances variation is approximately linear for a compression ratio between 10.2-16, with a power decrease of 2.5 % for each 5 degrees increase of the environment temperature; to be obtained a afterburning module with a 30 % reduction of the NO_x

reduction (at partial load) in comparison with the existing cogenerative power plant 2xST 18 – Suplacu de Barcau; to demonstrate the power increase and NO_x emissions reduction when injecting water in the intake device of TA2 gas turbine.

ACKNOWLEDGEMENTS

The research have been made based on contracts 22-097/2008, 22-108/2008 (Programme "Partnerships in priority fields") and 7N/2009, financed by Romanian Ministry of Education, Research, Youth and Sports. The consortium involved in contracts 22-097/2008, 22-108/2008 includes several companies in Romania: INCDT COMOTI Bucharest, SC OMV PETROM SA, UPB-CCT Bucharest, SC OVM-ICCPET SA Bucharest, SC ICEMENERG SA Bucharest, SC ERG SRL Cluj, SC TERMOCAD SRL Cluj.

REFERENCES

1. Technology characterization: Gas turbines, 2008, http://www.epa.gov/chp/documents/catalog_chptech_gas_turbines.pdf (accessed 17 July 2014).

2. Notiuni introductive privind ciclurile termodinamice ale motoarelor termice si turbinelor cu gaze. Diagrame entropice de stare, determinarea randamentelor termodinamice, http://www.rasfoiesc.com/inginerie/tehnica-mecanica/Notiuni-introductive-privind-c67.php (accessed 17 July 2014).

3. Contemporary Cogeneration Technologies, http://www.energymanagertraining.com/Journal/02112007/ContemporaryCogenerationTechnologies.pdf (accessed 25 June 2014).

4. Abam F., Ugot I., Igbong D., Performance analysis and components irreversibilities of a (25 MW) gas turbine power plant modeled with a spray cooler, American J. of Engineering and Applied Sciences 2012; 5 (1): 35-41, http://thescipub.

com/html/10.3844/ajeassp.2012.35.41 (accessed 17 June 2014).

5.   Pankaj P., Better power generation from gas turbine along with improved heat rate, International Conference–PowergenIntl 2000-Florida, Bangkok, ASME 2003-Atlanta, http://www.energymanagertraining.com/announcements/issue25/winners_papers_Issue25/05_PankajKPatel.pdf (accessed 17 June 2014).

6.   Hosseini R, Beshkani A., Soltani M., Performance improvement of gas turbines of Fars (Iran) combined cycle power plant by intake air cooling using a media evaporative cooler, Energy Conversion and Management 2007; 48: 1055–1064,

7.   Bagabir A., Khamaj J., Hassan A., Experimental and theoretical study of micro gas turbine performance augmentation, Emirates Journal for Engineering Research 2011; 16 (2): 79-88, http://www.engg.uaeu.ac.ae/ejer/issues/V16/pdf_iss2_16/7%20Ahmed%20Bagabir-22-22-F11_EXPERIMENTAL_Formatted%20_2__authors.pdf (accessed 22 June 2014).

8.   P. dos Santos, Andrade Cl., Zaparoli E., Comparison of different gas turbine inlet air cooling methods, World Academy of Science, Engineering and Technology 2012; 6, http://www.waset.org/publications/2686/comparison-of-different-gas-turbine-inlet-air-cooling-methods.pdf (accessed 17 June 2014).

9.   Etminan V., Moghiman M., Bajestan E., Boghrati M., Performance improvement of simple cycle gas turbine by using fogging system as intake air cooling system, http://www.google.ro/url?sa=t&rct=j&q=&esrc=s&source=web&cd=95&ved=0CGMQFjAEOFo&url=http%3A%2F%2Fprofdoc.um.ac.ir%2Farticles%2Fa%2F102366.doc&ei=lTpaU7KUFeOI7AbsjYHQDQ&usg=AFQjCNHCYaE5k7lhoIXTYF2D03ETtWw-ww (accessed 17 June 2014).

10.   Espanani1 R., Ebrahimi S., Ziaeimoghadam H., Efficiency Improvement Methods of Gas Turbine, Energy and

Environmental Engineering 2013; 1(2): 36-54, 2013 DOI: 10.13189/eee.2013.010202, http://www.hrpub.org/download/201309/eee.2013.010202.pdf (accessed 17 June 2014).

11.   Gopinath V., Navaneethakrishnan G., Performance Evaluation of Gas Turbine by Reducing the Inlet Air Temperature, International Journal of Technology Enhancements and Emerging Engineering Research 2013; 1 (1): 20-24, http://www.ijteee.org/final-print/aug2013/Performance-Evaluation-Of-Gas-Turbine-By-Reducing-The-Inlet-Air-Temperature.pdf (accessed 20 June 2014).

12.   Al-Doori W., Parametric Performance of Gas Turbine Power Plant with Effect Intercooler, Modern Applied Science 2011, 5 (3): 173-184

13.   Kyprianidis K., Future Aero Engine Design: An Evolving Vision, Advances in Gas Turbine Technology, Dr. Ernesto Benini (Ed.), InTech, 2011, ISBN: 978-953-307-611-9, DOI: 10.5772/19994. Available from: http://cdn.intechopen.com/pdfs-wm/22893.pdf (accessed 27 June 2014).

14.   Barbu E., Vilag V., Popescu J., Ionescu S., Ionescu A., Petcu R., Cuciumita C., Cretu M., Vilcu C., Prisecaru T, Afterburning Installation Integration into a Cogeneration Power Plant with Gas Turbine by Numerical and Experimental Analysis, Advances in Gas Turbine Technology, Dr. Ernesto Benini (Ed.), InTech, 2011, ISBN: 978-953-307-611-9, DOI: 10.5772/19994. Available from: http://www.intechopen.com/books/advances-in-gas-turbine-technology/afterburning-installation-integration-into-a-cogeneration-power-plant-with-gas-turbine-by-numerical-(accessed 17 June 2014).

15.   Barbu E., Petcu R., Vilag V., Silivestru V., Prisecaru T., Popescu J., Cuciumita C., Tomescu S., Gas Turbine Cogeneration Groups Flexibility to Classical and Alternative Gaseous Fuels Combustion, Progress in Gas Turbine Performance, Dr. Ernesto Benini (Ed.), InTech, 2013, ISBN: 978-953-51-1166-5, Available from: http://www.intechopen.com/books/progress-in-gas-turbine-performance/gas-turbine-cogeneration-

groups-flexibility-to-classical-and-alternative-gaseous-fuels-combustion (accessed 17 June 2014).

16. Motahar S., Alemrajabi A., Performance augmentation of an aero-engine gas turbine using steam injection, International Conference on Energy and Environment 2006 (ICEE 2006), http://www.iasj.net/iasj?func=fulltext&aId=57646 (accessed 27 June 2014).

17. Tyagi K., Khan M., Effect of gas turbine exhaust temperature, stack temperature and ambient temperature on overall efficiency of combine cycle power plant, International Journal of Engineering and Technology 2010; 2 (6): 427-429, http://www.enggjournals.com/ijet/docs/IJET10-02-06-18.pdf (accessed 17 June 2014).

18. Begovic M., Maintaining declared performance in gas turbines during increased ambient temperatures, Energija 2009; 58 (2): 192-207.

19. Kurz R., Gas turbine performance, Procedings of the thirthy-fourth turbomachinery Symposium, 2005, 131-146, http://turbolab.tamu.edu/proc/turboproc/T34/t34-14.pdf (accessed 17 June 2014).

20. Pavri R., Moore G., Gas turbine emissions and control, http://site.ge-energy.com/prod_serv/products/tech_docs/en/downloads/ger4211.pdf (accessed 27 June 2014).

21. Homji C., Mee Th., Gas turbine power augmentation by fogging of inlet air, http://turbolab.tamu.edu/proc/turboproc/T28/Vol28010.pdf (accessed 27 June 2014).

22. Kumar P., Krishna A., Raju G., Comparative Analysis on Performance of a Gas Turbine Power Plant with a Spray Cooler, International Journal of scientific research and management (IJSRM) 2013; 1 (7): 354-358.

23. Ionel I., Ungureanu C., Popescu F. Analiza nivelului de emisii poluante prin schimbarea combustibilui la cuptoarele de tratament termic. http://www.tehnicainstalatiilor.ro/articole/images/nr12_76-82.pdf (accessed June 14, 2014).

24. 24. Al-dawoodi M., Theoretical study for the effect of different fuels on the perfomance of open gas turbine power plant, http://www.google.ro/url?sa=t&rct=j&q=&esrc=s&source =web&cd=119&ved=0ciybebywcdhu&url=http%3a%2f %2fwww.uobabylon.edu.iq%2fpublications%2fapplied_ edition4%2fpaper_ed4_32.doc&ei=3unau5halir07abm_yh4 aq&usg=afqjcnho5qxani118jnp9p733fwu7ki5uq (accessed 17 June 2014).

25. Anderson L., Gas turbine design for cogeneration balancing efficiency and reliability for maximum economy, http:// www.gasnet.com.br/conteudo/11938/Gas-turbine-design- for-cogeneration-balancing-efficiency-and-reliability-for- maximum-economy (accessed 27 June 2014).

26. Brun K., Kurz R., Gas turbine life limiting rffects on inlet and interstage water injection, Procedings of the thirthy-fourth turbomachinery Symposium 2005, 45-52, http://turbolab. tamu.edu/proc/turboproc/T34/t34-06.pdf (accessed 27 June 2014).

27. Ibrahim Th., Rahman M., Alla A., Study on the effective parameter of gas turbine model with intercooled compression process, Scientific Research and Essays 2010; 5(23): 3760-3770, http://www.academicjournals.org/article/ article1380620575_Ibrahim%20et%20al.pdf (accessed 27 June 2014).

28. Dharap A., Ghanwat A., Inlet Air cooling for Gas turbines, Air Conditioning and Refrigeration Journal, October-December 1999, http://www.ishrae.in/journals/1999oct/article01.html (accessed 27 June 2014).

29. Suryan1 A. et all., Experimental investigations on impaction pin nozzles for inlet fogging system, Journal of Mechanical Science and Technology 2011; 25 (4): 839-845.

30. Makky A., Spray modeling tutorial using ANSYS-CFX, 2012, http://www2.warwick.ac.uk/fac/sci/eng/study/pg/students/ esrhaw/spray_modelling_using_ansys.pdf (accessed 27 June 2014).

31.   Wilson C.D., Riggins D.W. Performance characterization of turboshaft engines: work potential and second-law analysis. American Helicopter Society 58th Annual Forum. Montreal, Canada, June 11-13. 2002.

32.   Boyce M. P., Gas Turbine Engineering Handbook, Fourth Ed., Butterworth-Heinemann, 2012, ISBN 978-0-12-383842-1.

33.   Pimsner V., Maşini cu palete (Bladed machines). Ed. Tehnică, Bucharest, 1988.

34.   Carlanescu C., Manea I., Ion C., Sterie St. (1998). Turbomotoare – Fenomenologia producerii si controlul noxelor, Editura Academiei Tehnice Militare, Bucuresti

35.   Fagbenle R., (2010). Exergy and Environmental Considerations in Gas Turbine Technology and Applications, Gas Turbines, Gurrappa Injeti (Ed.), ISBN: 978-953-307-146-6, InTech, DOI: 10.5772/10206. Available from: http://www.intechopen. com/books/gas-turbines/exergy-and-environmental-considerations-in-gas-turbine-technology-and-applications (accessed 27 June 2014).

36.   Caresana F, et all, (2010). Micro Gas Turbines, Gas Turbines, Gurrappa Injeti (Ed.), ISBN: 978-953-307-146-6, InTech, DOI: 10.5772/10211. Available from: http://www.intechopen. com/books/gas-turbines/micro-gas-turbines-mgts-(accessed 27 June 2014).

37.   Henkel N, Schmid E., Gobrecht E., Operational flexibility enhancements of combined cycle power plants, 2007, https://www.cee.siemens.com/web/ua/ru/production_ energy/Documents/Service_Operational%20Flexibility%20 Enhancements%20of%20Combined%20Cycle%20 Power%20Plants.pdf (accessed 27 June 2014).

38.   Barbu E., Fetea G., Petcu R., Vilag V., Dragasanu L., Afterburning installation of 2xST 18 cogeneration power plant – investigations on combustion and NOx emissions, Chemical Engineering Transactions 2013; 34: 37-42, DOI:10.3303/ CET1334007, 2013.

4

Computational Fluid Dynamics Analysis of Greenhouses with Artificial Heat Tube

Nuno Couto[1], Abel Rouboa[1,2], Eliseu Monteiro[1], and José Viera[3]

[1]Department of Engineering, Universidade de Trás-os-Montes e Alto Douro, Vila Real, Portugal

[2]Department of Mechanical Engineering and Applied Science, University of Pennsylvania, Philadelphia, USA

[3]Research Department of Agriculture Secretary, Lisbon, Portugal

ABSTRACT

With the workmanship decrease in farms, the necessity to rationalize the use of other inputs and the development of technology has rapidly expanded the use of computer simulation in agricultural systems. One of the agricultural systems in which the modeling process of plant growth has been more engaged is the greenhouse production for horticultural crops. In Mediterranean climate, it is during the night that the energy losses are important and can be compensated with an artificial heat input. In this work an experiment was performed in a greenhouse in the north of Portugal. Temperature values in several points and air velocity in the aperture were measured during the night for three different cases: natural convective heating (case A); artificial heating tubes (AHT) (case B); AHT and natural ventilation (case C). A CFD simulation, carried out using FLOTRAN module of ANSYS, was also performed in two-dimensional configuration to obtain the indoor air temperature and velocity fields for the three cases. A very good agreement between experimental and numerical temperature values were verified, which allows to validate the adopted numerical procedure. In case A, the average temperature was 2.2°C. An average increase of 6.7°C and 3.5°C on the air temperature was obtained for the case B and case C, respectively. These results clearly emphasis the influence of each thermal load on greenhouse indoor air properties.

INTRODUCTION

In contrast to most agricultural production systems, greenhouses enable the growth and productivity of crops to be manipulated by controlling the indoor climate. A greenhouse is a closed space surrounded by translucent walls allowing better internal environmental conditions than the natural ones. In fact, greenhouses are widely used in the whole world due to their low cost. The airflow, the temperature and humidity vary strongly inside these closed spaces and depend of the outside climate. The temperature inside the greenhouses is controlled by heating systems. These heating equipments depend partially on the fuel source available as natural gas, fuel oil, diesel, or kerosene. Natural gas is the cleanest of the petroleum fuels and the most efficient to use without cost effective [1]. The exhaust gases from all heating systems contain combustion products that can cause damages on plants seedlings. Other products cause similar damage by emitting sulphur dioxide and nitric oxide. Several researchers are trying to optimize fertilization process to solve environmental problems.

Temperature, humidity, CO_2, solar radiation and air velocity determines the greenhouse microclimate. These variables are affected by the external weather, by the properties of the greenhouse cover, and by the properties of the plants.

In Portugal the greenhouses are covered by polyethylene rather than glass, therefore the heat losses by radiation at night could be considerable, especially in conditions of cloudless sky [2].

In Mediterranean climate, the radiation frost is dominant and is responsible for significant damages in horticulture crops during winter and earlier spring. This occurs often in Portugal because, generally, greenhouses do not have heating systems.

Simulation of environmental conditions in greenhouses is a powerful tool in modern horticulture. The heat ventilation process is the driving force for the air circulation and temperature distribution in greenhouses [3].

Several studies on natural ventilation were based on estimations of a global air exchange rate [4] and simulations of air temperature and a global vegetation temperature using a big leaf model [5,6] and energy balance methods [7].

In all these methods several heat transfer coefficients are assumed empirically and they are not able to clearly mapping airflow patterns and temperature profiles.

Meanwhile, some progress in flow modelling by computational fluid dynamics (CFD), in air circulation and temperature measurements, has recently been made for a closed greenhouse [4] and in a two-span naturally ventilated greenhouse [8]. Up till now, very few studies detailed climate in full-scale greenhouses heated by Artificial Heat Tube (AHT) have been conducted by both experiments and modeling.

In this work an experiment was performed in a polyethylene covered greenhouse in night conditions in the north of Portugal, and for three different cases:

- Case A: natural convective heating (no heater and no natural ventilation effect);
- Case B: AHT—artificial heating tubes (with heater and no natural ventilation effect);
- Case C: AHT and natural ventilation (heater and natural ventilation effect).

The measured temperature values should indicate the influence of each thermal load in the indoor air properties and allows the validation of several numerical procedures. The FLOTRAN module of ANSYS is used to obtain the indoor air temperature and velocity fields. This module performs CFD simulations using the finite element method, therefore a coupled differential equations system are solved, which means that it is only needed to validate one of the air properties to accept as reliable all the others.

MATERIALS AND METHODS

Experimental Set up

The experimental greenhouse (22 m length, 8.5 m width and 3.5 m height), located in the north of Portugal, is cover with a polyethylene sheet and have traditional discontinuous vent openings on the top. A schematic view of the experimental greenhouse is shown in Figure 1.

Temperature at several positions along the paths was measured by thin thermocouples. The coordinates of the nodes for five paths are shown in Figure 1. This will be of great significance later on the result analysis.

Since this region is characterized by a predominant northerly wind, channeled by a small valley, symmetric airflow was assumed along the direction with respect to each opening.

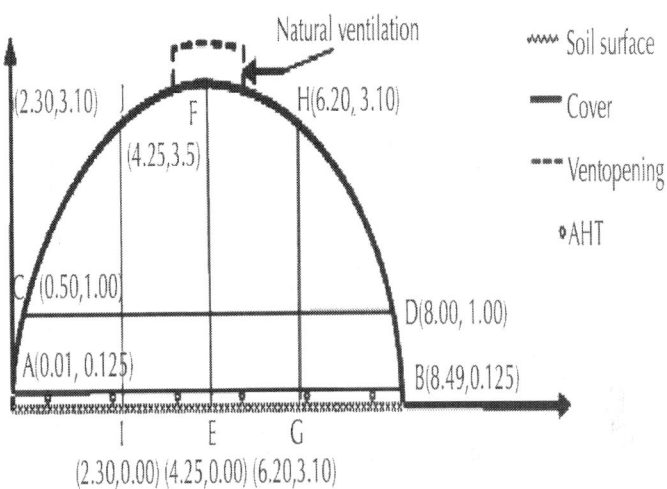

Figure 1: Schematic design of the experimental greenhouse and coordinates of nodes.

Therefore, a transversal section was selected to explore the flow patterns and the external air speed and temperature were measured by two three-dimensional sonic anemometers.

It is during the night that the energy losses are important and can be compensated with an artificial heat input. On the other hand, in this type of climate, the energetic contribution from solar radiation usually avoids heating during the day. During sunny days, it is usually necessary to ventilate greenhouses for limiting temperature elevations, and therefore to prevent the prejudicial limit (biological maximum) to be reached.

The classical heating systems are designed exclusively for the greenhouse aerial ambient. Although, the plant growth is intensively affected by its aerial environment, some physiological problems in winter cultivations are caused by inappropriate soil thermal conditions. These conditions can decrease the nutrient absorption process. On the other hand, if the greenhouse floor is heated, it will become into a source of heat to the environment. Therefore, in our experiment the heating system is composed by polyethylene tubes arranged on the floor into which hot water flows from a hot spring originated from geothermal aquifer.

Mathematical Model

The energy exchange between greenhouse and environment are complex, they can be summarized by:

- Thermal radiation from the soil, atmosphere, greenhouse environment and vegetation, emitted through greenhouse structure and cover;
- Natural convection of the indoor air;
- Forced convection caused by wind flow;
- Conduction in soil and cover.

The greenhouse indoor air flow is turbulent, therefore an adequate mathematical model avoiding the use of empirical heat transfer coefficients should be performed. Incompressible Reynolds averaged Navier-Stokes equations with the k-e model is used, also because it provides adequate CPU (Computing Process Unit) time and residual values [9].

This model was implemented in FLOTRAN module of ANSYS[a]. According to these variables, the continuity equation, momentum conservation equations, turbulent and dissipated energy (k-e) conservation equations, for an incompressible fluid in Cartesian coordinates, are written in conservative form as [10]:

$$DivU = 0 \tag{1}$$

$$\frac{\partial U}{\partial t} + U \cdot \nabla U + \nabla p \pm \left(v + c_u \frac{k^2}{\varepsilon} \right) \left(\nabla U + \nabla U' \right) = 0 \tag{2}$$

$$\frac{\partial \rho k}{\partial t} + \frac{\partial \rho V_x k}{\partial x} + \frac{\partial \rho V_y k}{\partial y}$$

$$= \frac{\partial\left(\frac{\mu_t}{\sigma_k}\frac{\partial k}{\partial x}\right)}{\partial x} + \frac{\partial\left(\frac{\mu_t}{\sigma_k}\frac{\partial k}{\partial y}\right)}{\partial y}$$

$$+ \mu_t \phi - \rho \varepsilon + \frac{C_4 \beta \mu_t}{\sigma_t}\left(g_x \frac{\partial T}{\partial x} + g_y \frac{\partial T}{\partial y}\right) \qquad (3)$$

$$\frac{\partial \rho \varepsilon}{\partial t} + \frac{\partial \rho V_x \varepsilon}{\partial x} + \frac{\partial \rho V_y \varepsilon}{\partial y}$$

$$= \frac{\partial\left(\frac{\mu_t}{\sigma_\varepsilon}\frac{\partial \varepsilon}{\partial x}\right)}{\partial x} + \frac{\partial\left(\frac{\mu_t}{\sigma_\varepsilon}\frac{\partial \varepsilon}{\partial y}\right)}{\partial y} + \mu_t \frac{\varepsilon}{k}\phi - C^2 \frac{\rho \varepsilon^2}{k}$$

$$+ \frac{C_1 \beta C_\mu (1 - C_3)\rho k}{\sigma_t}\left(g_x \frac{\partial T}{\partial x} + g_y \frac{\partial T}{\partial y}\right) \qquad (4)$$

The parameters used in k-ε model are presented in Table 1.

Resolution Method

The numerical method used by ANSYS[a] is based on Finite Element Method. The steady solution of the governing equations is given in each square element of the discretized whole domain. In order to solve the linear system, TDMA (Tri-Diagonal Matrix Algorithm) is used as solver [11]. Two-dimensional (V_x and V_y) velocities, pressure, turbulence kinetic energy (k) and turbulence kinetic energy dissipation ratio are a DOF (degrees of freedom) for each element. The convergence criteria of TDMA are 10^{-5} for the two component velocities V_x and V_y, and 10^{-3} for pressure, turbulence kinetic k and turbulence energy dissipation ratio e.

The ANSYS[a] module, FLOTRAN, allow solve easily the two-dimensional system equations cited above. The obtained solutions are pressure, temperature and velocity distribution in a single-phase.

Table 1: Modified k-ε model constants

Const	Modified k-e
C_1	0.43
C_2	1.9
C_m	0.09
C_s	1.0
σ_y	1.2
σ_t	1.0
C_3	1.0
C_4	0.0
b	0.0

For each element, ANSYS[a] code calculates velocity components, pressure, and temperature from the conservation of three properties: mass, momentum, and energy.

The requirements of meshes for turbulence model are more restrictive than those for laminar flow. Due to this fact, the "quad" element size has to be about 0.025 m. In the zone of higher gradients of temperature, velocities or pressure, in particular near the walls, the mesh size has to be refined with a factor of four as seen in Figure 2.

Boundary Conditions

In Table 2 are presented the boundary conditions (Dirichlet kind) used for each case.

- Development of the flow near outlet boundaries.

Stability Analysis

The convergence and numerical stability was obtained by observing the rate of change of the solution on the monitor and the behavior of relevant dependent variables during the iterations (Ite). Controlled variables were: velocity (V), pressure (P), temperature (T), and turbulence quantities such as kinetic energy (degree of freedom ENKE) and kinetic energy dissipation rate (ENDS).

The convergence monitors (CM) are a normalized measure of the solutions rate of change from iteration to iteration. Denoting by the general field variable, f, any DOF, the convergence monitor is defined as follows [9]:

$$CM = \frac{\sum_{i=1}^{n} \left| \phi_i^k - \phi_i^{k-1} \right|}{\sum_{i=1}^{n} \left| \phi_i^k \right|}$$

(5)

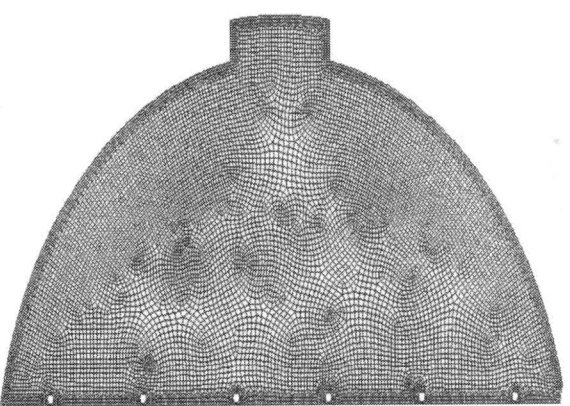

Figure 2: Element mesh of the 2-D model of the greenhouse.

Table 2: Boundary conditions

	Gravity(m/s²)	Temperature (°C)				Exterior air velocity (m/s)
		Floor	Cover	Heater	Exterior air	
A	9.81	10	−3	-	-	-
B	9.81	10	−3	60	-	-
C	9.81	10	−3	60	-3	1

Table 3: Convergence of our solution (CM values)

Case	Ite	C(h)	V_x	V_y	P	T	ENKE	ENDS
A	1.5×10^4	99	1×10^{-6}	1×10^{-6}	1×10^{-5}	1×10^{-7}	1×10^{-5}	1×10^{-5}
B	1.5×10^4	98	1×10^{-4}	1×10^{-4}	1×10^{-4}	1×10^{-6}	1×10^{-4}	1×10^{-3}
C	1.2×10^4	89	1×10^{-3}	1×10^{-3}	1×10^{-3}	1×10^{-5}	1×10^{-3}	1×10^{-2}

After initial variations, convergence monitors decrease as the analysis approaches convergence, and the number of iterations depends on several factors, such as:

- Complexity of the geometry;
- Mesh refinement;
- The turbulence level indicated by the Reynolds number.

Table 3 describes, for each case, the number of iterations (Ite), the computational time in hours (C(h)), and the residual for each variables (V_x, V_y, Press, ENKE and ENDS).

A special note should be made, as CPU time is concerned, that convergence took a relative long time to reach an acceptable solution. This indicates that the mesh is probably too refined or the TDMA solver was not the best choice.

RESULTS AND DISCUSSION

Model Validation

In this section, the experimental and numerical results of temperature are compared. As expected, the temperature distribution inside the greenhouse is affected for the three cases: close without heater tube, close with heater tube and open window with heater. The Temperature results are presented in relative form ($T_r = T - T_f$). T_r is the relative temperature, T_f is the floor temperature without heater (10°C). The presented Temperature for the case A (close greenhouse without heater) shows that the ambient temperature ranges from 1.7°C to 3°C. The left side of the greenhouse is colder than the right side on about 1.3°C. This is due to the wind orientation. In fact the wind is directed from right to the left side of the greenhouse. For the case B, where the heater is open and the greenhouse is close, the temperature increase from 8.3°C to 10.5°C. For the greenhouse with a heater tube and opening windows a similar airflow pattern was observed as in case B but with higher velocity values and lower temperature. Indeed, the opening windows allow the freeze air entrance with a velocity equal to 1 m/s. In this case, a temperature of the greenhouse fluctuates from 4.9°C and 7.4°C.

As shown in Figure 3, the comparison between the experimental data and the numerical results present a very good agreement for all of the cases. As it's shown, the average error between experimental and numerical results are 0.2°C. In case B, where the heater is on, the error on the temperature values are higher than in the two other cases.

TEMPERATURE AND VELOCITY RESULTS

Figure 4 shows the temperature and velocity profiles numerically obtained for each case and for the path AB (represented in Figure

1). In this figure, the temperature is more or less constant between heater tubes locations, and the airflow velocity is higher in the middle of the path. The temperature values are higher than presented above, this is due to path location been only 10cm from heat tubes. On the zone where the heater tube is located the temperature value is around 55°C. The velocity values vary along the path AB between 0.05 m/s to 0.4 m/s. The high values are observed in the middle of the path. This is due to greenhouse geometry and wind orientation.

For the three cases and on the path CD, the temperature and velocity profiles numerically obtained are shown on Figure 5.

Is observed that the relative temperature is approximately constant, and the airflow velocity is fluctuate between 0.4 m/s on the left side and 0.1 m/s near the first heater tube location.

The difference in the average temperature value between cases B and C allows verifying that ventilation provokes sufficient energy loss to decrease the temperature in 3°C. The airflow velocity, in case C is higher than in the other two cases, but only in a certain part of the path. This is due to the fact of the exterior air flow (right to left direction), which feeds the descending convection flow.

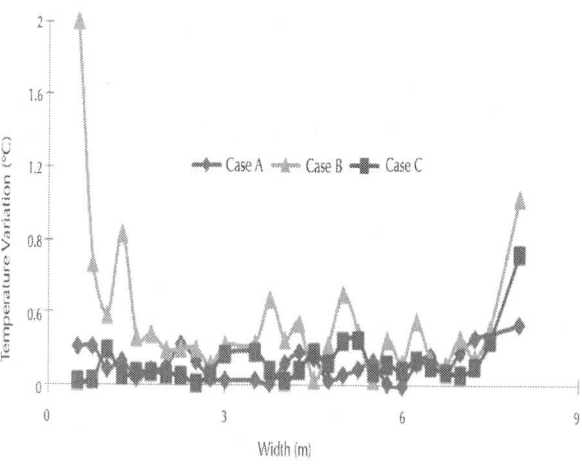

Figure 3: Comparison between experimental and numerical results of temperature for the 3 cases.

In Figure 5 is clearly showed the influence of the AHT and natural ventilation in temperature (about 7°C in case B and about 3°C in case C). This influence is almost insignificant in the airflow velocity profile.

As far as temperature is concerned a special note should be made for the thermal inversion that happens in all the vertical paths but is more visible in case B. This behaviour has its cause in the AHT system set at 0.125 m from the soil and the wind orientation.

Air Velocity inside Greenhouse

In Figures 6-8 the velocity vectors of fluid flow are represented, respectively in situations A, B and C.

The turbulent regime is lower in a greenhouse without heating and no ventilation. Figure 6 shows a main convection flow in anti-clockwise.

The polyethylene heating tubes change substantially the turbulent regime inside the greenhouse.

In this situation, behind the main convection flow, it is possible to identify six other flows generated by the heating tubes.

The turbulent regime increases even further in the situation in which there is air flux between exterior and interior areas. In this situation it is possible to identify secondary convection flows. Another important observation is the fact that, the descending cold airflow velocity is greater than in the situations without air exchange with the exterior, caused by the addition of cold exterior air.

CONCLUSIONS

In the literature, many researchers are working on the analysis of the wind effect on ventilation [3-5,12], but no work has been done on the wind effect for the greenhouse with AHT implanted in the soil. The influence of AHT in the temperature and air velocity was examined numerically. Three cases were studied, the closed greenhouse without heater tubes, the heater tubes implanted in the soil in closed greenhouse and finally, the second case with opening windows.

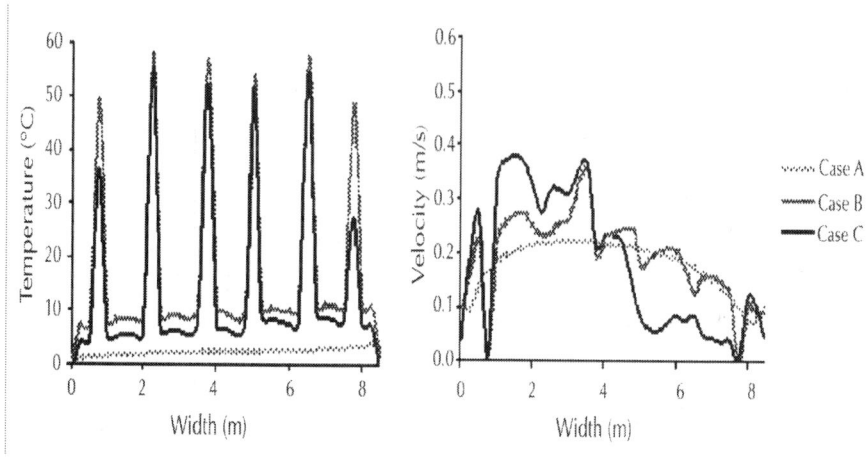

Figure 4: Temperature and velocity profiles on path AB.

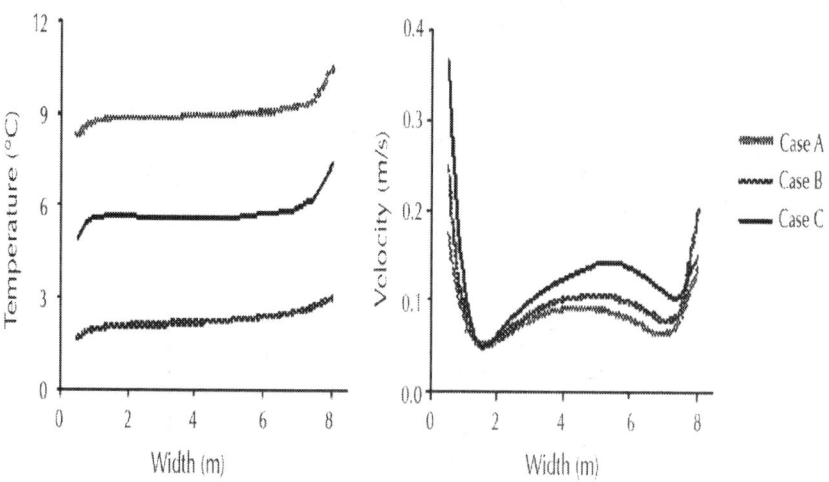

Figure 5: Temperature and velocity profiles on path CD.

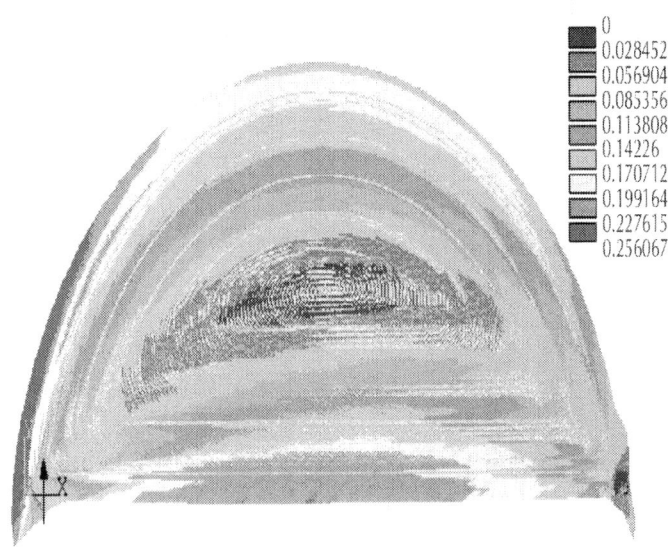

Figure 6: Vector of velocity for case A using FLOTRAN.

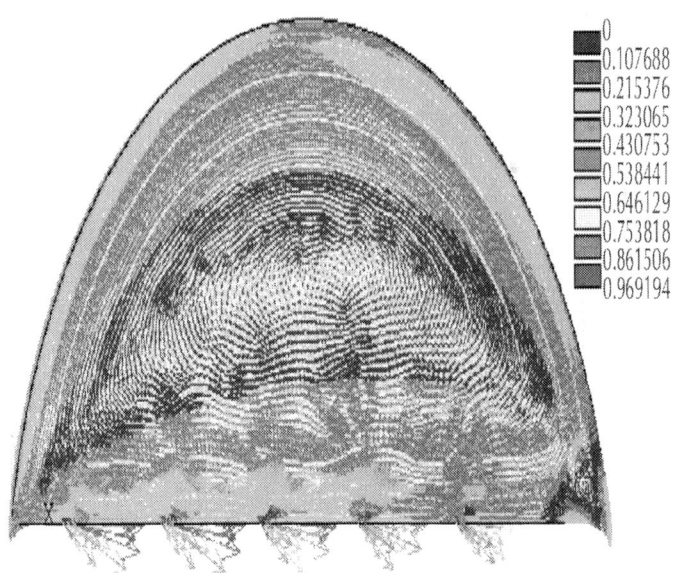

Figure 7: Vector of velocity for case B using FLOTRAN.

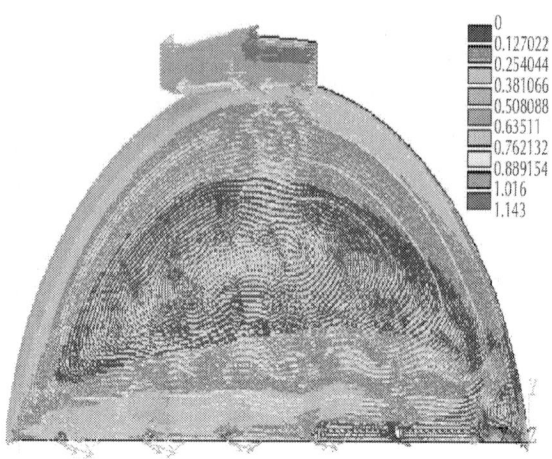

Figure 8: Vector of velocity for case C using FLOTRAN.

Temperatures obtained numerically were compared with experimental data. These demonstrate a good concordance and allow to validate the used numerical code to simulate heat and mass transfer in this studied domain. Then temperatures distribution in a horizontal plane situated 1.125 m from the ground are presented. For each case, the distribution of temperature inside the greenhouse was quite different and the resultant temperature profile was mainly affected by airflow. When the wind enters with a velocity equal to 1 m/s and a temperature equal to 1°C, the temperature inside the greenhouse decrease significantly (from 19°C to 16°C). This decrease depends deeply not only on the wind velocity and its temperature but also on its direction, as studied.

Air velocity distribution along the greenhouse presents a main circulation in the middle of the greenhouse for all situations. In respect to the openings, both air velocity and temperature had a uniform distribution along the greenhouse and air velocity varied between 0.3 and 0.4 m/s. When air flow was parallel to the openings and each opening acted as an inlet and an outlet, we observed regions inside the greenhouse, mainly in the middle of greenhouse, with very low air velocities (0.05 - 0.1 m/s). Consequently, temperature gradually increased between the two openings up to

5°C higher than the outside air.

This paper describes and evaluates the computational facilities using the finite element method to study the effects of heating tubes and natural ventilation on greenhouses indoor air properties especially during the night.

In opposite to earlier works, usually based on thermal loads, in this study the incompressible Reynolds averaged Navier-Stokes equations with the k-e model was performed. This numerical procedure avoids the use of empirical heat transfer coefficients and provides adequate CPU (Computational Processing Unit) time and residual values. This mathematical model was implemented in FLOTRAN module of ANSYS[a], which is based on finite element method. Good agreement has been observed between the numerical and experimental values. This allows to validate the Computational Fluid Dynamics code used in this work.

Results shown that the heating tubes increases the temperature in about 6.7°C. If both heating tubes and natural ventilation are introduced this increase reduces to about 3.5°C. Turbulent regime is lower in case A, and it increases slightly when the heating system is introduced (case B), and it increases significantly in case C due to the effect of natural ventilation.

The simulation of these processes using ANSYS can be a good path to explore, namely in the simulation of three dimension resolution and optimizing the size of the element mesh in order to reduce the computation time.

REFERENCES

1.   I. Seginer, "Optimal Control of the Greenhouse Environment: An Overview," Acta Horticulturae, Vol. 406, 1996 pp. 191-201.

2.   O. Korner and H. Chala, "Design for Improved Temperature Integration Concept in Greenhouse Cultivation," Computers and Electronics in Agriculture, Vol. 39, No. 1, 2003, pp. 39-59. doi:10.1016/S0168-1699(03)00006-1

3.   T. Boulard and B. Draoui, "Calibration and Validation of a Greenhouse Climate Control Model," Acta Horticulturae, Vol. 406, 1996, pp. 49-61.

4.   B. J. Bailey and W. Day, "The Use of Models in Greenhouse Environmental Control," Acta Horticulturae, Vol. 491, 1999, pp. 93-99.

5.   T. Boulard and S. Wang, "Experimental and Numerical Studies on the Heterogeneity of Crop Transpiration in a Plastic Tunnel," Computers and Electronics in Agriculture, Vol. 34, No. 1-3, 2002, pp. 173-190. doi:10.1016/S0168-1699(01)00186-7

6.   S. Wang and T. Boulard, "Measurement and Modelling of Radioactive Heterogeneity in a Greenhouse Tunnel," Acta Horticulturae, Vol. 534, 2000, pp. 139-146.

7.   S. Wang and J. Deltour, "Theoritical Study of Natural Ventilation Flux in a Single Span Greenhouse," Biotechnologie, Agronomie, Société et Environnement, Vol. 2, No. 4, 1998, pp. 256-263.

8.   S. Wang, J. Pieters and J. Deltour "Studies on Radiometric, Thermal and Climatic Properties of a New Greenhouse Covering Materials," Acta Horticulturae, Vol. 491, 1998, pp. 324-333.

9.   J. J. Costa, L. A. Oliveira and D. Blay, "Test of Several Versions for the K-E Type Turbulence Modelling of Internal Mixed Convection Flows," International Journal of Heat and Mass Transfer, Vol. 42, No. 23, 1999, pp. 4391- 4409. doi:10.1016/S0017-9310(99)00075-7

10.   J. H. Ferziger and M. Perić, "Computational Methods for Fluid Dynamics," 2nd Edition, Springer Verlag, New York, 1999. doi:10.1007/978-3-642-98037-4

11.   J. C. Tannehill, D. A. Anderson and R. H. Pletcher, "Computational Fluid Mechanics and Heat Transfer," 2nd Edition, Taylor & Francis Ltd., Oxfordshire, 1997.

12.   B. J. Bailey, "Optimal Control of Dioxide Enrichment in Ventilated Greenhouses. Workshop on Management," Identification and Control of Agriculture Buildings,

Universidade de Trás-os-Montes e Alto Douro, Vila Real, 1998, pp. 1-15.

5

A Fundamental Equation of Thermodynamics that Embraces Electrical and Magnetic Potentials

Salama Abdel-Hady

Department of Mechanical Engineering, CIC, Cairo, Egypt.

ABSTRACT

This paper introduces an energy-analysis of the flow of electrical charges and magnetic flux in addition to the flow of heat into a thermodynamic system. The analysis depends on the results of a held experiment on a magnet that attracted iron balls and a Faraday's discovery as well as similarities between the laws characterizing the flow of electric charges, magnetic flux and heat. As the electric charge and magnetic flux produce entropy changes in some applications as magnetic hysteresis and capacitive deionization, we may express the electric charges and magnetic flux in terms of entropy changes times their corresponding potentials by analogy to expressing heat as a product of temperature and entropy changes. Introducing such postulates into the held energy-analysis; a new fundamental equation of thermodynamics that embraces the flow of electric charges and magnetic flux in terms of the electric and magnetic potentials was derived. The derived equation proved its truth in numerous studies of general energy interactions. Such postulates help also in defining the electric charge and magnetic flux as categories of electromagnetic waves of assigned electric or magnetic potentials and offer plausible explanations of newly discovered phenomena in the fields of electromagnetism and thermodynamics as Kerr effect and magnetic expansion.

INTRODUCTION

A simple experiment was run on a control mass that encloses a permanent magnet immersed in a water basin. The magnet was allowed to attract iron-balls moving steadily toward the magnet along an inclined plane. The results of such experiment show a reduction in the temperature of the water. Such decrease in temperature or internal energy of the magnet and the system is considered as a result of a work done by the magnet on the moving balls. So, it is possible to postulate the magnetic flux as a form of energy or electromagnetic waves similar to the heat emitted from a hot body. Reviewing Faraday's discovery when inserting dielectric slabs between the plates of a condenser, the energy stored in the capacitor circuit decreased by the same ratio of increase of the electric charge on the condenser plates [1]. Such result proves also that the electric charge transferred to the dielectrics increases by the same ratio the stored energy of the inserted dielectrics.

According to Maxwell's comments regarding his equations [2], electricity and magnetism were supposed to be of a wave-like nature as the electromagnetic waves since an oscillating electric field generates an oscillating magnetic field and the oscillating magnetic field in turn generates an oscillating electric field, and so on. In former researches [3,4], it was discussed also common features of the magnetic flux and electric charges. In this article, it is reviewed the similarities between laws characterizing electric charges, magnetic flux and heat flow to prove, analytically, their identity as special categories of electromagnetic waves. The processing of substances in magnetic refrigeration cycles led to a defined description of the produced magnetic-hysteresis in terms of increase of the system's entropy and temperature [5]. Similarly, thermodynamic work required to separate solutions during capacitive deionization processes is fully expressed in terms of decrease of the system's entropy [6]. According to such conclusions; it is possible to postulate the magnetic flux and electric charges in terms of entropy change times a corresponding potential by analogy to heat flux which is expressed in terms of temperature, as a thermal or heat potential, and the corresponding change of entropy. Introducing such postulates into the first law of thermodynamics led to a fundamental equation that embrace electric charge and magnetic flux as mechanisms of energy transfer and that proved its "trueness".

Many references defined the time's arrow as a property of entropy alone [7]. According to such definition, the time coordinate in Maxwell Equations and solutions was fuzzily substituted by entropy. Accordingly; the areas bounded by the oscillating electrical and magnetic potentials in the electromagnetic waves could represent the electrical and magnetic energy imparted by such electromagnetic waves. Such approach helped in postulating the electric charges and magnetic flux as special categories of electromagnetic waves of assigned potentials. Such postulates offer many plausible explanations for phenomena in the fields of thermodynamics and electromagnetism [8] as the magnetic expansion [9], Kerr effect [10] and the phenomena of discharge of clouds, colours of charges and magnetic dipole moments.

EXPERIMENTAL ANALYSIS OF THE MAGNETIC FLUX

As an approach to prove similarity of the heat emitted from a hot body and the magnetic flux emitted from a magnet; a permanent magnet was immersed in an insulted water basin and was allowed to attract polished iron balls of along an inclined smooth glass plane, Figure 1. The temperature of water was recorded. As the magnet performed a mechanical work, a decrease in the water temperature was measured by a digital thermometer. By applying the first law of thermodynamics on the magnet-water closed-system [11], we get the following equation:

$$C_F \frac{\Delta t}{\Delta \tau} = n \; m_b \; g \; s \; \sin\varphi$$

(1)

where C_F is the flask's heat capacity, n is the number of attracted balls along the inclined plane per unit time, m_b is the mass of any iron ball, g is the gravity acceleration, s is the length of the path of the irons ball along the inclined smooth plate; j is angle of inclination of the smooth plate and Δt is the change of temperature during the time interval "Δt".

(Experimental data: thermal capacity of the flask = 1.26 kJ/deg, mass of each iron ball = 0.3 kg, number of attracted balls per minute = 42 balls, s = 20 and = 30 deg.)

Figure 2 shows the results as calculated from Equation (1) and as recorded during the attraction process. As a conclusion, there is a loss of the magnet's internal energy during attraction of the iron balls. Such energy was substituted by decrease in the temperature or internal energy of the system. Such conclusion proves the behavior of a magnet emitting magnetic flux is similar to a hot body emitting heat or electromagnetic radiation. So, the magnetic flux may be considered as a form of energy or electromagnetic waves.

ANALYSIS OF ELECTRIC FLUX

According to the Faraday's discovery [1], when the space between the capacitor plates was filled by a dielectric material, the capacitance of the capacitor increased by a dielectric constant k defined as follows:

$$C = \kappa \; C_{air} \tag{2}$$

Faraday found also that the addition of such dielectric material between the plates increased the initial charge on the plates of a capacitor, q_i to a final value q_f defined by the equation [1]:

$$q_f = \kappa \; q_i \tag{3}$$

By comparing the initial and final energies stored in the capacitor's circuit, U_i and U_f, it was found that the initial energy decreased by the same coefficient κ according to the equation [12]:

$$U_f = \frac{U_i}{\kappa} \tag{4}$$

According to Equations (3) and (4); the coefficient   equals the ratios of increase of the electric charge on the capacitor plates and the decrease of the internal energy of the capacitor circuit due to insertion of the dielectric material. Such result shows that the loss of the stored energy in the capacitor circuit is absorbed as an electric charge on the capacitor plates. According to the energy – conservation principles; the electric charge can be considered as a form of energy that is transferred to or from the dielectrics and causes the simultaneous increase of the capacitor charge and decrease of the internal energy of its circuit.

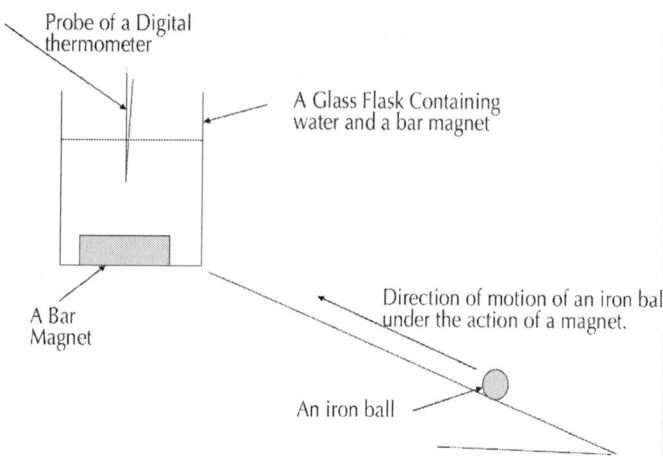

Figure 1: Measurement of magnet's work during attraction of iron balls.

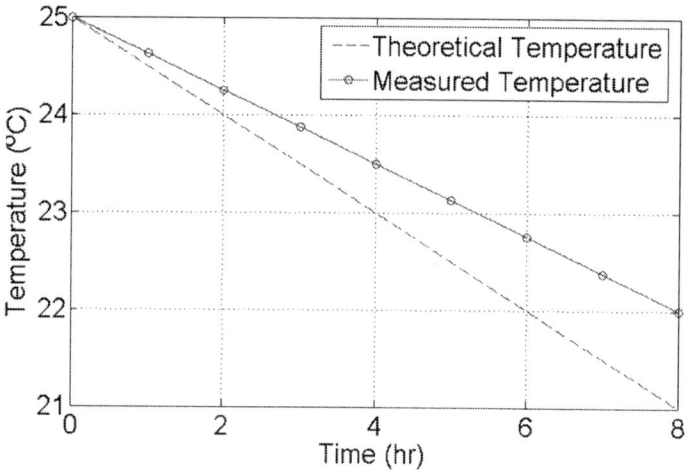

Figure 2: Measurement and calculation results of the system temperature.

MAXWELL WAVES

Regarding the origin of the electromagnetic wave equations; Maxwell combined displacement current with some of the other equations of electromagnetism and obtained a second-order partial differential equation that describes the propagation of electromagnetic waves through free space or a medium. The homogeneous form of such equation is written in terms of the electric field E and the magnetic field B as follows [12]:

$$(\nabla^2 - \frac{1}{c^2}\frac{\partial^2}{t^2})E = 0$$

$$(5)$$

$$(\nabla^2 - \frac{1}{c^2}\frac{\partial^2}{t^2})B = 0$$

$$(6)$$

where c is the speed of light in the free space (c = 299,792,458 m/second), and t is the time.

The general solution of such electromagnetic wave equation is a linear superposition of waves of the form [13]

$$E(r,t) = g(\varphi(r,t)) = g(\omega t - k.r)$$

$$(7)$$

$$B(r,t) = g(\varphi(r,t)) = g(\omega t - k.r)$$

$$(8)$$

Virtually, both forms represent a well-behaved function "g" of dimensionless argument "j" where

w: is the angular frequency (in radians per second), and k: is the wave vector (in radians per meter).

In addition, for a valid solution, the wave vector and the angular frequency are not independent; they must adhere to the dispersion relation [13]:

$$k = |k| = \frac{\omega}{c} = \frac{2\pi}{\lambda}$$

$$(9)$$

where: k is the wave number and λ is the wavelength.

Maxwell commented the results of his solutions as follows; the agreement of the results seems to show that light and magnetism are affections of the same substance, and that light is an electromagnetic disturbance propagated through the field according to electromagnetic laws [12]. Such comment considers magnetism of the same substance as electromagnetic waves. Similarly, Maxwell derived a wave form of the electric and magnetic equations, revealing a wave-like nature of electric and magnetic fields. According to Maxwell [13], electricity and magnetism were considered of a wave-like nature as electromagnetic waves, since an oscillating electric field generates an oscillating magnetic field and the magnetic field in turn generates an oscillating electric field, and so on. Finally; according to Maxwell comments electric charge and magnetic flux may be considered also as forms of electromagnetic waves or energy as they are components of such waves.

SIMILARITY OF FLUXES

Energy is restricted in this research to the heat radiation propagating as electromagnetic waves or, according to quantum mechanics, as photons. Similarities between such forms of energy as heat, electric charge and magnetic flux are primarily found into their common quantization and conservation properties. The energy is generally quantized into photons of quanta v, where h is the Planck's constant [13,] which is conserved according to principles of energy conservation. The electric charge is quantized and conserved too. The elementary charge of a single electron or proton is found experimentally as follows [12]:

$e = 1.60 \times 10^{-19}$ Coulomb

It is found that any positive or negative charge q show has a quantized value that is expressed as follows [12];

$$q = ne, \text{ where } n = \pm 1, \pm 2, \pm 3,\dots (n \in Z) \quad (10)$$

The fact that electric charge is quantized may be considered as one of the most profound mysteries of the physical world. However, considering the charge as a form of energy, which are naturally quantized, removes such mystery. Similarly, it is found that the charge is conserved as the net charge in a system or object is preserved too. The magnetic flux is also quantized into quanta expressed generally as [12]:

$$\Phi_e = n\frac{h}{2e}, \text{where } n = \pm1, n = \pm2, n = \pm3 \quad (n \in Z)$$

(11)

It is also found that many laws characterizing energy transfer, as heat, are similar to the laws that characterize the transfer of electric charges and the magnetic flux. The heat transfer by conduction is expressed as [14]:

$$\frac{\dot{Q}}{A} = \frac{k}{\Delta x}(T_1 - T_2) = \frac{1}{k/\Delta x}(T_1 - T_2)$$

(12)

Similarly, the charge transferred is expressed by Ohm's law as [11]:

$$I = \frac{\dot{Q}_e}{A} = \frac{1}{R}(V_1 - V_2)$$

(13)

Equations (12) and (13) show an analogy between the energy transfer by heat and electricity. Heat transfer in the first equation is driven by the difference in temperature and in the second equation the charge transfer is driven by the difference in electrical potentials. However, the laws concerning the transfer of magnetic flux are not yet formulated, but there is an approach to its formulation as a magnetic flux emergence in the sun is directly traced on the solar surface (in visible-white light) by the presence of dark, mainly round-shaped areas, called sunspots, surrounded by brighter regions of higher magnetic potential called plagues [15]. Measurements of magnetic fields at the solar surface shows that sunspots tend to be grouped in pairs, one with positive and one with negative magnetic polarity which may be characterized by flow of magnetic flux according to similar equations as Equations (12) and (13). Such similarities between heat, electric charge and magnetic flux prove they are forms of energy or electromagnetic waves.

A FUNDAMENTAL ENERGY EQUATION

The first and second laws of thermodynamics may lead to a fundamental equation of thermodynamics in the form [14]

$$dU = T \, dS - p \, dV + E \, de + \sigma \, d\Omega + \sum_i \mu_i dn_i$$

(14)

where:

dU: Change in internal energy of a T.D. system;

T dS: Thermal Energy imparted to a system represented as the product of the temperature and the change of entropy;

P dV: work of expansion of volume (dV) under the pressure p;

σ dW: energy imparted by new surface area formation dW at the surface tension σ;

$\sum \mu_i \, dn_i$: work of chemical change dn_i at given chemical potential μ_i.

However, Equation (14) does not include the magnetic energy that may be transferred to any system. In addition, its R.H.S. includes the term "E de" that corresponds to the electric energy imparted to the system. This term assumes the parameters "E" and "e" are properties that define the state of a general thermodynamic system and de is an exact differential or a differential of a property. According to former analysis, the electric charges "e" transferred to the system represent energy in transfer of a similar nature as heat and it should not be considered as a property of a thermodynamic system. So, Equation (15) is incorrect according to the principles of thermodynamics.

Introducing the electric charge, magnetic flux and heat flux as mechanisms of energy transfer to a thermodynamic system; a modified form of Equation (14) can be written as follows:

$$dU = \delta Q_{th} + \delta Q_{elect} + \delta \Phi_{mag} - p \, dV + \sigma \, d\Omega + \sum_i \mu_i dn_i$$

(15)

The processing of substances in magnetic refrigeration cycles leads to a defined description of the produced magnetic hysteresis in terms of an entropy increase and the system temperature [4]. So, temperature and entropy changes were actually achievable by the flow of magnetic flux into a system. In this context, the fact that the working material displays hysteresis [16], calls for a better physical understanding of the role of irreversibility in the magnetization process. Such understanding can be based on describing the magnetic flux by similar terms as heat flux, i.e. in terms of magnetic potential times a corresponding entropy increase. Such representation removes the ambiguity found in Bill queries [9]. Similarly; revising the work of Biesheuvel in the thermodynamic cycle analysis of capacitive deionization, [6], where the thermodynamic work required in solutions separation during capacitive deionization processes was fully utilized to decrease the entropy of the system. Such analysis implements also the expression of the electric charge in terms of the product of electric potential times the corresponding entropy increase.

According to the second law of T.D., the flow of heat into a system generates an entropy increase of such system expressed by the relation [10]:

$$Q_{thermal} = \int T \, dS_{thermal}$$

(16)

According to the previous analysis and the derived similarities, the electric charge can be expressed by a similar expression as follows.

$$Q_{elect} = \int E \, dS_{electric}$$

(17)

where Q_{elect} is electric charge imparted to a system, E is the electrical field intensity or potential, in analogy to temperature as the thermal potential, and $dS_{electric}$ is the entropy increase associated by such transferred electric energy or charge.

According to modern literatures [12], the term "B" represents the magnetic flux density. However, such density can be assumed proportional to the magnetic field intensity, H, according to the following relation (in free space) [12]:

$$B = \mu_0 H \tag{18}$$

where μ_0 ($= 4\varpi\ 10^{-7}$) is called permeability of free space or magnetic space constant. Such direct proportionality allows introducing the magnetic field intensity to replace the magnetic flux density in Maxwell's solution which is analogous to the electric field intensity.

So, the magnetic flux can be expressed as follows:

$$\Phi_{mag} = \int H \ dS_{magnetic} \tag{19}$$

where Φ_{mag} is the magnetic flux imparted to a system, H is the magnetic field intensity and $dS_{magnetic}$ is the generated entropy by the transferred magnetic flux. By substituting Equations (16), (17) and (19) into Equation (14); a modified fundamental equation of thermodynamics can be expressed as follows:

$$dU = T\ (dS)_{th} + E\ (dS)_{el} + H\ (dS)_{mag}$$
$$-p\ dV + \sigma\ d\Omega + \sum \mu_i dn_i \tag{20}$$

where $T\ (dS)_{th}$, $E\ (dS)_{el}$ and $H\ (dS)_{mag}$ correspond to the thermal, electric and magnetic energies imparted to the system, respectively, and $(dS)_{th}$, $(dS)_{el}$ and $(dS)_{mag}$ represent the entropy productions by the corresponding transferred thermal, electric and magnetic energies. The main corollary of such fundamental equations is the possibility of representing the electric charges transferred or imparted to a system into an E-s diagram and the magnetic flux imparted to a system into an H-s diagram similar to the T-s diagram that represent the heat transferred to such system.

MAXWELL EQUATIONS

The arrow of time is found as property of entropy alone [7]. Accordingly; time in Maxwell Equations (5) and (6) may be replaced by entropy. Such transformation leads to modified solutions of Maxwell equations that may be stated as follows:

$$E(r,t) = s(\varphi(r,s))$$

(21)

$$H(r,t) = s(\varphi(r,s))$$

(22)

Accordingly, the electromagnetic waves can be represented in E-s and B-s planes by replacing the time axis in Figure 3 by entropy. In such representation; the areas enclosed by the oscillating fields, electric or magnetic, represents the imparted electric and magnetic energies as postulated in Equations (17) and (19).

Searching for a plausible answer of a frequently asked question: If the electromagnetic waves are considered, according to the quantum mechanics, as photons; then, how the photon can have a charge or a magnetic polarity [14]? The answer implements a modified solution of Maxwell Equations which is not identical to that stated by Equations (7) and (8). In the following solution it is assumed a non-zero initial electrical potential "DE_i" in the electric component of the electromagnetic wave. Hence, the following solution may satisfy the postulated Maxwell's Equations (21) and (22), where the time is replaced by entropy:

$$E(r,t) = s(\varphi(r,s)) +/- \Delta \bar{E}$$

(23)

$$H(r,t) = s(\varphi(r,s))$$

(24)

Such solution assumes some categories of electromagnetic waves whose electric potential is oscillating around a specified value of non-zero electric potential but its magnetic potential is oscillating about a zero magnetic potential. The graphical representation of such solution is seen in Figures 4 and 5, where Figure 4 represents a positive charge and Figure 5 represents a negative charge.

The net charge in each imparted sinusoidal wave has a value which may be positive or negative according to the sign of the charge found by the following integral in a complete wave:

$$Q_{elect} = \int_0^{2\pi} \bar{E}\, dS_{elect}$$

(25)

The given integration may represent an electric charge of a positive or negative potentiality according to the direction of the wave oscillations if it is around positive or negative values of $\Delta\bar{E}$.

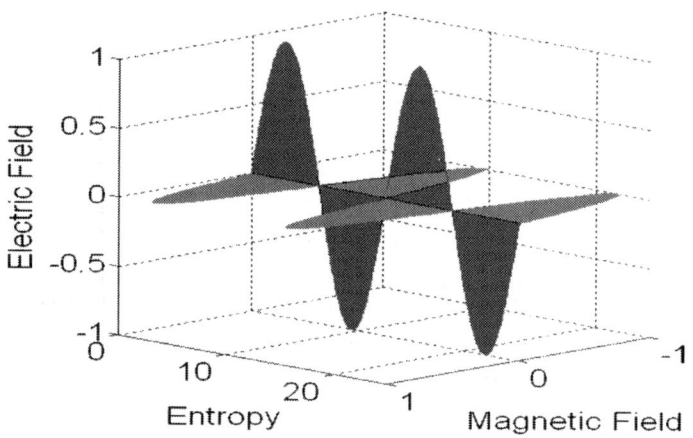

Figure 3: Electromagnetic waves in E-s and H-s planes.

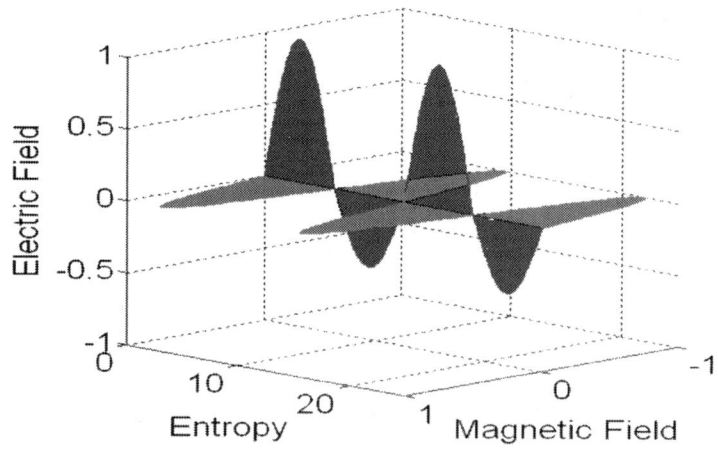

Figure 4: Graphical representation of a positive electric charge.

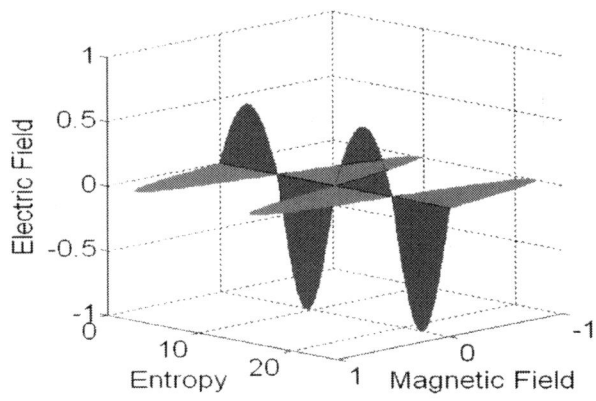

Figure 5: Graphical representation of a negative electric charge.

Similarly, it is assumed in the following solution a non-zero initial magnetic potential "$\Delta\bar{H}$" in the magnetic component of the electromagnetic wave. So, the following modified Maxwell's solution may represent flow of magnetic flux:

$$H(r,t) = s(\varphi(r,t)) +/- \Delta\bar{H}$$

(26)

$$E(r,t) = s(\varphi(r,t))$$

(27)

Such solution is represented graphically in Figure 6 where the net bounded area has a non-zero magnetic potential. So, the net magnetic flux in each imparted sinusoidal wave has the value:

$$\Phi = \int_0^{2\pi} \bar{H} \, dS_{mag}$$

(28)

So, the postulated solutions of Maxwell Equations and the introduced definitions of electric charges and heat flux assign definite potentials for such fluxes in terms of ($\Delta\bar{E}$) and ($\Delta\bar{H}$), with definite polarity or direction. The units of the electric charge or the magnetic flux are stated as forms of energy in Joules. Hence the units of dS_{elect} and dS_{mag} will be in Joule/Volt and Joule/Oersted respectively in analogy to the units of $dS_{thermal}$ in Joule / K [6]. Such definitions remove confusions in the units of magnetism and electricity.

REVIEWING THE INTRODUCED FUNDAMENTAL EQUATION

As previously mentioned, the definitions of the electric charge and magnetic flux, as modified forms of EM waves, can be directly concluded from Maxwell's statements and the similarity of the laws governing such entities. However, such definitions introduce also plausible explanations to many discovered phenomena. The definition of the electric charge as electromagnetic waves explains the behavior of the cathode rays as waves in the famous Crookes tube [15], where the cathode rays, defined previously as flow of electric charges or electrons, travelled in straight lines, produce a shadow when obstructed by objects and the rays could pass through thin metal foils without disturbing them (Tested by New Zealander Ernest Rutherford using gold foil [15]). Hertz maintained it also was a wave. Hence, the introduced definition of electric charge as a flow of modified EM waves offer more plausible explanations of the found characteristics of the Cathode rays as a form of electromagnetic waves.

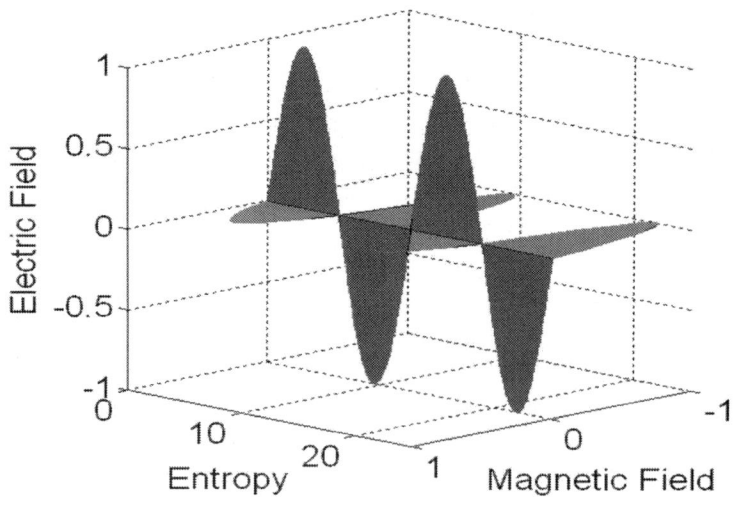

Figure 6: Graphical representation of magnetic flux.

Similarly, the introduced definitions of the electric charges and the magnetic flux as special categories of electromagnetic waves find plausible explanations to the "Kerr Effect" [9]. Such phenomena detect effects of electric and magnetic fields on the flow of electromagnetic waves. In case of electric fields, a quadratic electro-optic effect (QEO effect) is detected. In case of magnetic fields, a magneto-optic Kerr effect (MOK effect) is detected. The QEO effect is identified by a change in the refractive index of all materials when they are crossed by electromagnet waves in the presence of an applied AC or DC electric fields. Similarly; MOKE effect may be recognized, in some cases, by rotation of the plane of polarization of the transmitted light under the influence of magnetic field [9]. According to the introduced definitions, the light gains by the applied electric or magnetic fields the potentials differences($\Delta \overline{E}$) or ($\Delta \overline{H}$), that converts the light waves into electric charges or magnetic flux of different index of refraction or rotated plane of polarization.

Replacing the time axis by an entropy axis in the Maxwell's solutions and considering the propagating EM waves as propagations of electric and magnetic energies of limited entropy production and energy quanta proves the description of photons as tiny bundle of electric and magnetic fields [18].

Introducing entropy, as a probability measure of disorder [18], to replace the time in Maxwell equations lead to a probabilistic Maxwell's solution that is analogous to probabilistic wave solution $\Psi(x, t)$ of the Schrodinger's equation. The introduced fundamental equation can be valuable in studies that apply Poisson-NernstPlanck theory to investigate ion permeation and related transport processes [19].

Assigning a value of a definite electric potential for the charge, according to the stated definitions of electric charges, is similar to the assigned thermal potential, temperature, for the heat. Such electrical potential explains the phenomena of colored charges where each charge may have a different potential [8] and explain the phenomena of charging of clouds [12].

CONCLUSIONS

Reviewing experimentally and analytically the similarity and analogy between heat, electric charges and magnetic flux, the electric charge and magnetic flux were proved as special categories of energy or modified forms electromagnetic waves that have electrical or magnetic potentials. Previous analysis served also in expressing the magnetic flux and electric charges in terms of entropy change times a corresponding potential by analogy to heat flux. Introducing such postulates into the first law of thermodynamics led to a modified fundamental equation that embrace electric charge and magnetic flux as mechanisms of energy transfer. Considering time's arrow as a property of entropy alone [7], the time in the Maxwell equation is replaced by entropy such that the areas bounded by the oscillating electrical and magnetic potentials represent the electrical and magnetic energies imparted by the electromagnetic wave. Such representation helps in expressing the qualities or potentials for the electric charge and the magnetic flux in analogy to the heat quality (temperature). The offered postulates lead to modified definitions of electric charge and magnetic flux in addition to plausible explanations of newly discovered phenomena in the fields of electromagnetism and thermodynamics as Kerr effect and magnetic expansion.

REFERENCES

1.   D. T. Ryan, "Toward a cognitive-historical understanding of Michael Faraday's research: Editor's introduction," Perspectives on Science, Vol. 14, No. 1, 2006.

2.   J. D. Jackson, "Classical electrodynamics," 3rd Ed., Wiley, New York, 1998.

3.   S. Abdelhady, "Thermodynamic analysis of electric charges and magnetic flux," Cairo 11th International Conference on Energy and Environment, Ghurgada, pp. 175–185, March 2009.

4.   S. Abdelhady, "A fuzzy approach to the physics of electromagnetic waves and atomic particles," Cairo 10th International Conference on Energy and Environment, Luxor, pp. 234–241, March 2007.

5.   V. Basso, et al., "Effect of material hysteresis in magnetic refrigeration cycles," International Journal of Refrigeration, Vol. 29, No. 8, pp. 1358–1365, December 2006.

6.   P. M. Biesheuvel, "Thermodynamic cycle analysis for capacitive deionization," Journal of Colloid and Interface

7.   Science, Vol. 332, No. 1, pp. 258–264, April 2009.

8.   J. J. Halliwell, J. Pérez-Mercader, and W. H. Zurek, "Physical origins of time asymmetry," Cambridge University Press, London, 1994.

9.   A. E. Shabad and V. V. Usov, "Electric field of a pointlike charge in a strong magnetic field and ground state of a hydrogenlike atom," Physical Review D, Vol. 77, No. 2, 2008.

10.   F. X. Hu, et al., "Magnetoresistances and magnetic entropy changes associated with negative lattice expansions in $NaZn_{13}$-type compounds LaFeCoSi," Chinese Physics, Vol. 14, No. 11, pp. 2329–2334, 2005.

11.   T. J. Englert, B. H. Chowdhury, and E. A. Grigsby, "Laboratory investigation of the electro-optic Kerr effect for the detection of transmission line faults," IEEE Transactions on Power Delivery, Vol. 6, No. 3, pp. 979–988, 1991.

12.   A. C. Yunus and A. B. Michael, "Thermodynamics: An engineering approach," McGraw-Hill Science Engineering, 2006.

13.   D. Haaiday, R. Resnick, and J. Walker, "Fundamentals of physics," 7th Ed., John Wiley & Sons, New York, 2004.

14.   C. F. Stevens, "The six core theories of modern physics," MIT Press, London, 1965.

15.   H. B. Callen, "Thermodynamics and an introduction to themostatistics," John Wiley & Sons, New York, 1985.

16. P. K. Shukla, et al., "Equivalent electric charge of photons in an electron – positron plasma," Physica Scripta, Vol. 62, No. 2–3, 2000.

17. M. N. O. Sadiku, "Elements of electromagnetics," Oxford University Press, Oxford, 2006.

18. B. Benny, "How do you reconcile EM fields with frequency of light?" April 2009. http://www.physicsforums. com/showthread.php?p=2161363

19. D. J. Griffiths, "Introduction to quantum mechanics," Benjamin Cummings Publishing Company, San Francisco, 2004.

20. S. A. Zhou and M. Uesaka, "Modeling of transport phenomena of ions and polarizable molecules: A generalized Poisson–Nernst–Planck theory," International Journal of Engineering Science, Vol. 44, No. 13–14, pp. 938–948, 2006.

Chapter 6

Grand Potential Formalism of Interfacial Thermodynamics for Critical Nucleus

Atsushi Mori and Yoshihisa Suzuki

Institute of Technology and Science, the University of Tokushima, Tokushima, Japan

ABSTRACT

In nucleation theories, the work of formation of a nucleus is often denoted by W = ΔG. This convention misleads that the nucleation should be considered in the isothermal-isobaric system. However, the pressure in the system with a nucleus is no longer uniform due to Laplace's equation. Instead, the chemical potential is uniform throughout the system for the critical nucleus. Therefore, one can consider the nucleation in the grand ensemble properly. Accordingly, W is found to be the grand potential difference and the interfacial tension is also turned to be an interfacial excess grand potential. This treatment is not entirely new; however, to explicitly treat in the grand potential formalism is for the first time. We have successfully given an overwhelmingly clear description.

INTRODUCTION

The work of formation of a nucleus is often written as ΔG. It leads one to understand the work of formation of the critical nucleus as a difference of the Gibbs energy. The meaning of the form of the work of formation of a critical nucleus (Equation (4) in the text) becomes, however, clear straightforwardly if we deal the system including a critical nucleus as an isothermal-isochoric open system. The treatment as an isothermal-isobaric closed system brings confusions. The concept of the Gibbs dividing surface is more clearly understood in the isothermal-isochoric open system. As will be stated in the text, the treatments of an isothermal-isochoric open system appeared in literatures already. In this paper, we will give a clearer and direct statement in the grand potential formalism for nucleation, aiming at helping researchers who are not specialists in thermodynamics. In other words, by describing with definite terminologies we will put forward understandings— some terminologies will be for the first time used definitely in this paper.

Gibbs established the interfacial thermodynamic formula for the work of formation of a critical nucleus in 1870s [1]. Since then, this subject was sometimes revisited and developed and/or extended [2-21]. One of true developments may be descriptions for the curvaturedependence of the interfacial tension [4,22-31]; as shall be described in Section 1.2, the interfacial tension γ is assumed to be known prior to the calculation of the radius R of the nucleus in the Gibbs formula. In other words, Gibbs' treatment (Section 1.2) alone does work for evaluating the work of formation of the critical cluster if the interfacial tension is independent of the curvature of the interface. Later Tolman's treatment was extended to the binary system [32]. Clarifying the meaning of the Gibbs dividing surface as done previously [2,3,5,11] and shall be done in Section 1.3 is helpful for general readers to avoid confusions, but not entirely new. Also embodiment of the variation of area A by defining the conical system with the solid angle ω around the center of the nucleus, such as done previously [2, 3,5,9,11,21], is, indeed, very helpful for ones who need rigorous arguments, but also not entirely new.

Throughout this paper we restrict ourselves to the case of spherical interfaces for simplicity and for the sake of avoiding complexity for better understanding. For example, two principal curvatures appear in general; this may bring confusion. Also, for the same sake we limit ourselves to unary cases. Also, for the same sake we omit the structure of both two phases; if at least one of the coexisting phases is crystalline, the interfacial tension becomes, strictly speaking, crystallographic orientation dependent.

Issue

One of purposes of the thermodynamics of nucleation is to calculate the reversible work of formation of a critical nucleus of a stable phase in an undercooled parent phase [1]. Through this work, W, one can obtain the steady-state nucleation rate as $J_s = J_o e^{-W/k_B T}$ with $k_B T$ being the temperature multiplied by Boltzmann's constant. Not only in textbooks [34-36] but also in advanced research papers [14,37-45] the following expression (or essentially equivalent one) is seen for the work of formation of a critical nucleus:

$$\Delta G = n\Delta\mu + \gamma A,$$

(1)

where $\Delta\mu \equiv \mu^\beta - \mu^\alpha (<0)$ is the difference between the chemical potentials of the nucleating phase (β phase) and the parent phase (α

phase). The direct interpretation of Equation (1) is as follows. Limiting ourselves to the case that the molecular volumes (volumes per molecule) of the α and β phases are equal1, let us denote the molecular volume V_m. Hereafter, the subscript m indicates the molecular quantities. Then, the quantity n is defined as the number of molecule consisting the nucleus, which is equal to $4\pi R^3 / 3V_m = V^\beta / V_m$ with $V^\beta = 4\pi R^3 / 3$ being the volume of the nucleus. The first term in Equation (1) is the volume term, which is the reversible work associated with the transformation from the α phase to the β phase of n molecules. The second term in Equation (1) is the surface term, which is the reversible work to form a surface of area $A = 4\pi R^2$. Here, R is the radius of the nucleus; the rigorous definition of R will be given later. Remembering that the chemical potential is equal to the molecular Gibbs energy, the expression of ΔG seems at apparent appropriate. The question arises whether the expression of Equation (1) is only valid for the case that no volume change is associated with the α-β phase transition or not. Exact expression for the reversible work W was already given and the approximation which reduces the exact expression to Equation (1) was derived [11]. Also the expression of **ΔG** makes one understood at apparent that the interfacial tension γ is defined as the superficial interfacial Gibbs energy; also exact expression for γ was already given [7,11]. Unfortunately, the previous derivations were not so transparent. A clearer interpretation will be given in this paper in a framework of the grand potential formalism. This paper aims at leading the readers to a clear understanding of the work of formation of a nucleus and solving the misunderstanding. The meaning of the interfacial free energy (or the interfacial tension) γ becomes also clear; the interfacial tesion γ can be understood as the superficial grand potential [3-5,9,11,12].

Gibbs Interfacial Thermodynamics

To review the Gibbs' formalism for evaluating W is not only heuristic but also ingredient for understanding the thermodynamic "ensemble" appropriate for the system of nucleation. In other words, due to this one can find why the grand potential formalism is appropriate; that is, constant μVT condition is imposed. It is sufficient to limit ourselves to the unary case; formulation for the multi-component system is seen, for example, in a previous paper [46].

Consider a spherical nucleus of the β phase in an undercooled α phase of the chemical potential $\mu^\alpha = \mu$ at the temperature T The chemical potential μ and temperature T are regarded as those of the reservoir. Along with the isothermal condition, for the critical nucleus one can regard a cluster of the β phase is in equilibrium with the α phase with respect to the material transport. One can select $\{T,V,\mu\}$ as independent variables specifying the total system with V being the volume of the total system. The following is the procedure of the calculation of the work of formation of a critical nucleus.

1. The pressure of the β phase is determined by the equilibrium equation with respect to the materials transport, i.e.

$$\mu^\beta\left(p^\beta,T\right)=\mu.$$
(2)

2. Presuming the interfacial tension γ as known, the radius R is determined by Laplace's equation,

$$p^\beta - p^\alpha = \frac{2\gamma}{R},$$
(3)

    where p^α is the pressure of the α phase corresponding to (T,μ).

3. The work of formation of the critical nucleus of radius R is calculated by

$$W = -\left(p^\beta - p^\alpha\right)V^\beta + \gamma A.$$
(4)

We note that eliminating $p^\beta - p^\alpha$ using Laplace's equation (Equation (3)), Equation (4) is rewritten into

$$W = \frac{1}{3}\gamma A.$$

(5)

We should note that the work of formation of a critical nucleus consists of two terms; as has been mentioned the first term is the volume term and, in tern, the second term is the surface term. The first term in Equation (4) is the work to replace the α phase of volume V^β with the β phase. The second term, γ^A, is understood as the work associated with the formation of area A of the surface free energy γ per unit area. In other words, in writing the work of formation of the critical nucleus we divide the process of nucleus formation into two. One is to form a hypothetical nucleus of the β phase possessing the bulk properties throughout the entire volume V^β in the parent α phase. The other is regarded to that to form a actual structure of the interface.

Gibbs Dividing Surface and Surface of Tension

For the first one of the two works of formation of a critical nucleus, the mathematical surface of radius R is a key concept. This surface is called the Gibbs dividing surface. Owing to introducing the dividing surface one can divide the work of formation of a nucleus into two. The volume term is the work of formation of a hypothetical cluster as illustrated in Figure 1. The surface term of the form of γ^A is, however, not very general; this form is valid only for the surface of tension, which will be explained later. The general form includes a curvature-dependent term [7]. There are varieties of choices of the dividing surface. Most straightforward one is the equimolar surface; the total numbers of molecules of the hypothetical system and the real one are the same thereby. The dividing surface introduced in Section 1.2 is called the surface of tension as mentioned there. By this choice, the coefficient γ appears in the surface term in the work of formation of a critical nucleus coincides with the interfacial tension. The definition of the surface of tension is implicit; the choice so that the curvature-dependent term vanishes is the definition. For the choice of the surface of tension, Laplace's equation (Equation (3)) holds; Laplace's equation is the equation of the mechanical balance at the curved interface possessing the mechanical tension γ.

order parameter

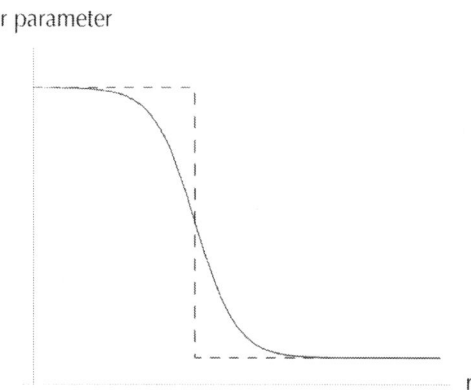

Figure 1: A schematic illustration of the profile of the order parameter (the density in, e.g., vapor-liquid case) with the horizontal axis indicating the distance from the center of the nucleus. In general, the order parameter varies between two bulk values gradually. Dashed lines indicate the hypothetical system, in which inside the dividing surface, indicated by a vertical dashed line, is occupied with a bulk β phase and outside with a bulk α phase.

Therefore, the interfacial free energy γ is called the interfacial tension.

WORK OF FORMATION OF CRITICAL NUCLEUS

Sometimes very unnatural variables are specified [2, 11]. That is, the internal energy E, the entropy S, and the amount of substances are selected as independent variable. The mass as well as the number of molecule can be employed as the amount of substances. Nevertheless, Nishioka [11, 13] derived a correct conclusion that γ is equal to the superficial grand potential through an entangled argument.

As pointed in Section 1.2 the chemical potential throughout the system is uniform. Along with the fact that the system is considered as isothermal, it is appropriate to select the temperature T and the chemical potential μ as independent variables. In this case, because at least one extensive variable is necessary for complete description, the total system volume V must be, in general, selected as one of the independent variables. We note that the uniformity of the chemical potential was already pointed out [2]; the treatment there was, however, not fully satisfactory.

Isothermal-Isochoric Open System and Grand Potential

As mentioned above the temperature and the chemical potential are uniform throughout the system. One can regard that the system is exposed to the isobaric reserver because if the chemical potential and the temperature are kept constant, the corresponding pressure, which is a function of T and μ, is also constant. In Figure 2 we illustrate an isobaric closed system and an isochoric open system; whereas in the former the system size changes after the nucleation, in the latter the system size is unchanged thereafter. Therefore, we should take into NPT v.s. μVT N, P, T

$$\mu(P, T), V, T$$

account the change of the total volume in calculation of the work of formation of a nucleus for the former case.

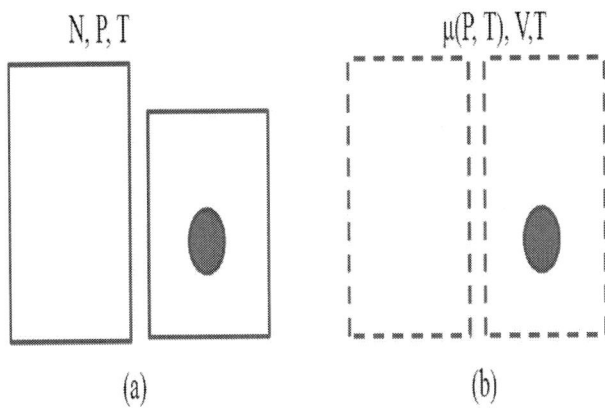

Figure 2: Comparison of the isobaric closed system and the isochoric open system before and after the nucleation. For clarity we assume that the nucleating phase is more condensed than the parent phase. In the isobaric case the total volume varies due to the nucleation. Accordingly, to figure out the work of formation of the nucleus in the isobaric system is somewhat complicated.

This is somewhat complicated. Hence, it is convenient to treat the system as isothermal-isochoric open one. Of course, two ways of description are both correct. The reversible work calculated as the Gibbs energy difference should coincide to that calculated as the grand potential difference. Indeed, a consideration with confusions led to the correct answer [33]. Unfortunately, in [33] the volume term and the surface term had been intertwined with each other; the form of Equation (5) has been eventually obtained.

At least in Japan, a thermodynamics class does not teach the grand potential systematically. One can, however, obtain isochoric open system by Legendre transformation of the isothermal-isochoric closed system, i.e., the independent variable is transformed from the amount of substances to the chemical potential to obtain this system [47]. The thermodynamic potential is obtained from the Helmholtz energy F by extracting N_μ (remember that $\mu \equiv (\partial F / \partial N)_{T,V}$ is thermodynamic conjugate variable to n); that is,

$$\Omega = F - N\mu = F - G = -pV,$$

(6)

where $G = N_\mu$ is the Gibbs energy. To reach to the last expression we have used the definition $G = F + p^V$. One may be familiar with this form in the grand canonical ensemble (μ^{VT} ensemble) through the bridging relation in this ensemble [48]. The thermodynamic potential Ω is the grand potential. We note that the grand potential (or merely the symbol Ω) already appeared in a thermodynamic expression for the interface in literatures [20,25,28,31,42,49-51] and a textbook [48]. In addition, the grand potential Ω may be familiar in the fields of the density-functional theory.

By virtue of the last expression of Equation (6), we obtain the volume term of the work of formation of a critical nucleus, as the grand potential difference between the system including the hypothetical nucleus and the homogeneous α phase, as

$$\Delta\Omega = -\left(p^\beta - p^\alpha\right)V^\beta,$$

(7)

where p^α and p^β are the pressures of respective bulk phases; even though there is no bulk part of the β phase in reality such as for a small nucleus, the pressure p^β is well defined (through Equation (2)). Due to the positive interfacial tension between the α and β phases, the pressure p^β of the phase inside the dividing surface is greater than p^α (thermodynamic derivation of this relation will be given in Section 2.2). In this way, we have the first term in Equation (4), which is negative and corresponding to the volume bulk term in Equation (1).

Work of Formation of Critical Nucleus

As argued up to now, we know that the work of formation of a critical nucleus is composed of the volume term, which is corresponding to the first term in Equation (1) and given by Equation (7), and the surface term, which is corresponding to the second term in Equation

(1). If the equilibrium with respect to the materials transport holds between the parent phase and the nucleus, the pressure inside the nucleus, p^β, is obtained by solving

$$\mu^\alpha \left(T, p^\alpha\right) = \mu^\beta \left(T, p^\beta\right),$$

(8)

which corresponds to Equation (2) and consistent to the isothermal open system (μ^{VT} ensemble). Because the α phase is metastable and the β phase is the stable phase; that is,

$$\mu^\alpha \left(T, p^\alpha\right) > \mu^\beta \left(T, p^\alpha\right)$$

(9)

holds, one can derive $p^\beta > p^\alpha$. Recalling the GibbsDuhem relation $d\mu = -S_m dT + V_m dp$, we draw schematically the chemical potentials as functions of the pressure in Figure 3; the larger the slope is, the larger the molecular volume V_m is. In Figure 3(a), we illustrate μ^α and μ^β for a normal case $\left(V_m^\alpha > V_m^\beta\right)$. Because the α phase is metastable (Equation (9)), the location of p^α is in the side $p > p^{eq}$. Therefore, from Equation (8) one can find the location of p^β as illustrated in Figure 3(a). An illustration for an abnormal case $\left(V_m^\alpha > V_m^\beta\right)$ such as the case of water-ice phase transition is given in Figure 3(b). The interpretation is logically the same.

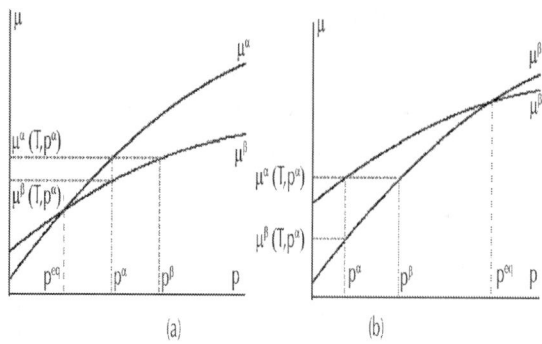

Figure 3: The μ-p relations are plotted for (a) a normal case $\left(V_m^\alpha > V_m^\beta\right)$ and (b) an abnormal case $\left(V_m^\alpha > V_m^\beta\right)$. Therefrom, one can confirm inequality $p^\beta > p^\alpha$.

In this way, the negativity of the volume term is understood. The criterion for the dividing surface has not been given yet. The surface term, in general, take a form [7,9,11]

$$\gamma\left(R\right)A + CdR.$$

(10)

Here, $\gamma(R)$ denotes that this coefficient depends on the criterion for the dividing surface. The surface of tension is defined by C(R)=0. Only for this choice of R, the coefficient $\gamma(R)$ coincide with the interfacial tension. In other words, the surface term consist of, in general, the interfacial area dependent term and the curvature dependent term. The surface of tension is defined for which the curvature dependent term vanishes. We note that $\gamma(R)$ takes the minimum for the surface of tension [7].

In this way, we have obtained Equation (4) for the work of formation of a critical nucleus. We give a note here. The work for the formation of the critical nucleus takes, however, the same value if the physical condition is unchanged; that is, it is not dependent on the criterion of the dividing surface. Therefrom, one can derive the relation between the general $\gamma(R)$ and the interfacial tension.

This was done by Kondo [7].

Noting $V^\beta = 4\pi R^3 / 3$ and $A = 4\pi R^2$, let us solve the equation that the derivative with respect to R of Equation (4) vanishes. By a simple calculation we have Laplace's equation (Equation (3)). This is a mechanical balance equation. Namely, in a case that two phases are coexisting via an interface of a curvature radius R with an interrfacial tension γ, the force acting from the inside of the interface due to the pressure p^β balances with the composed force of the force due to the outside pressure p^α and that due to the interfacial tension (corresponding to $p^\alpha + 2\gamma / R$). The quantity γ defined as the interfacial free energy per unit area of the interface is, if one chooses the surface of tension as the dividing surface, coincides with the mechanical interfacial tension. Readers can readily confirm the coincidence between the unit of the energy per area and the tension.

Now, let us derive the form of the first term in Equation (1), following Nishioka and Kusaka [13]. We start with the relation

$$\left(\frac{\partial \mu}{\partial p} \right)_T = V_m,$$

(11)

which is nothing other than the Gibbs-Duhem relation for the isothermal case. We consider a case that an incompressible β phase nucleus is nucleated in the α phase. Let us integrate Equation (11) for the β phase for p from p^α to p^β.

$$\mu^\beta \left(T, p^\beta \right) - \mu^\beta \left(T, p^\alpha \right) = \int_{p^\alpha}^{p^\beta} V_m^\beta dp = V_m^\beta \left(p^\beta - p^\alpha \right).$$

(12)

Eliminating $p^\beta - p^\alpha$ in Equation (4) using the equation derived by dividing Equation (12), we have an equation corresponding to Equation (1):

$$W = \frac{V^\beta}{V_m^\beta} \Delta\mu + \gamma A,$$

(13)

where

$$\Delta\mu = \mu^\beta \left(T, p^\alpha\right) - \mu^\beta \left(T, p^\beta\right)$$
$$= \mu^\beta \left(T, p^\alpha\right) - \mu^\alpha \left(T, p^\alpha\right).$$

(14)

To reach to the last expression, Equation (8) has been used. One can integrate Equation (11) for the α phase to obtain the form of Equation (1) in a case that the α phase is incompressible. This is, however, not the present concern. It should be noted that for a case that no volume change is associated with the $\alpha - \beta$ phase transition, a form far form Equation (1) is obtained [52], although in this case one has intuitively $\Delta(n\mu) = n\Delta\mu$ with $n = V^\beta / V_m^\beta = V^\beta / V_m^\alpha$.

GIBBS ADSORPTION ISOTHERM

In this section, we derive the Gibbs adsorption isotherm

$$\left(\frac{\partial\gamma}{\partial\mu}\right)_T = -\Gamma,$$

(15)

where μ represents the chemical potential of the materials reservoir, which is equal to μ^α, and Γ is the superficial number density per unit area of the interface, sometimes referred to as the excess number density or the interfacial adsorption quantity. A rigorous definition of Γ will be given later.

Conical System and Superficial Quantities

We define the system as a spherical cone as illustrated in Figure 4. In this definition, there are two variables describing the extent of the system; through the solid angle ω we can apply Euler's theorem for the homogeneous equation. Unlike previous papers [9,11,13], we define the system as open with the chemical potential. In those papers, the arguments were started with selecting the entropy S, the number μ of molecule N, the radius R_0, and the solid angle ω as independent variables. However, the argument becomes simplified with the selection of independent variables T and μ, instead of S and N. We note that R_0 is selected enough larger than R.

For the hypothetical system, because of the bulk properties, the following fundamental equations (Gibbs relations) hold for two parts of the system:

$$dE^{\alpha,\beta} = TdS^{\alpha,\beta} - p^{\alpha,\beta}dV^{\alpha,\beta} + \mu dN^{\alpha,\beta}.$$

(16)

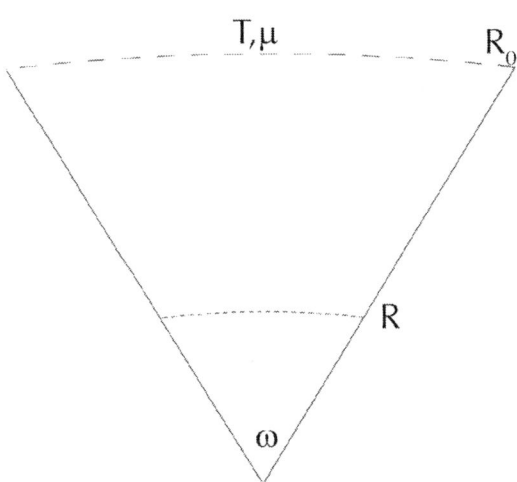

Figure 4: Conical system with the solid angle ω around the center of the nucleus. The system is defined as isochoric with the solid angle ω and the radius R_0. The system is exposed to the reservoir of the temperature T and the chemical potential μ.

Here, according to a convention E is used to represent the internal energy. This equation is rewritten in terms of the grand potentials $\Omega^{\alpha,\beta} = E^{\alpha,\beta} - TS^{\alpha,\beta} - \mu N^{\alpha,\beta}$ as

$$d\Omega^{\alpha,\beta} = -p^{\alpha,\beta}dV^{\alpha,\beta} - V^{\alpha,\beta}dp^{\alpha,\beta}$$
$$= -S^{\alpha,\beta}dT - p^{\alpha,\beta}dV^{\alpha,\beta} - N^{\alpha,\beta}d\mu \tag{17}$$

Those equations hold for both systems with the solid angle ω and the entire sphere $(\omega = 4\pi)$. In those expressions

$$V^{\beta} = \omega R^3 / 3, \tag{18}$$

$$V^{\alpha} = \omega \left(R_0^3 - R^3 \right)/3, \tag{19}$$

and we should note that R_0 and ω are independent variables.

Let us denote quantities for the entire spherical system by symbols with a superscript 4π and those for the system with the solid angle ω by symbols without a superscript. For a while, let us consider again a general dividing surface. Denoting the contribution due to the nucleus by..., the fundamental equation

$$d\Omega^{4\pi} = -S^{4\pi}dT - p^{\alpha}dV^{4\pi} - N^{4\pi}d\mu^{\alpha} + \cdots, \tag{20}$$

holds. Here, $V^{4\pi} = 4\pi R_0^3 / 3$ and because the R_0 is an independent variable, $dV^{4\pi} = 4\pi R_0^2 dR_0$. Let us rewrite Equation (20) using $S = (\omega / 4\pi)S^{4\pi}, V = (\omega / 4\pi)V^{4\pi}$ and

$N = (\omega / 4\pi)N^{4\pi}$. Because $dV = \omega R_0^2 dR_0 + (R_0^3 / 3)d\omega$ (from $V = \omega R_0^3 / 3$), we have

$$d\Omega = -SdT - p^{\alpha}\left[\omega R_0^2 dR_0 + \left(R_0^3 / 3 \right)d\omega \right] - Nd\mu^{\alpha}$$
$$= -SdT - p^{\alpha}\omega R_0^2 dR_0 - Nd\mu^{\alpha} + \sigma d\omega. \tag{21}$$

Here, we express the contribution of the nucleus by introducing the coefficient σ defined by

$$\sigma = \left(\partial \Omega/\partial \omega\right)_{T,R_0,\mu^\alpha} \left[=\left(\partial E/\partial \omega\right)_{S,R_0,\mu^\alpha}\right] \tag{22}$$

as previously done [2,3,5,9,11,13]. In those previous papers, the expression in the square brackets was given.

Differentiating $\Omega = (\omega/4\pi)\Omega^{4\pi}$ and using Equation (20), we have

$$d\Omega = \left(\omega/4\pi\right)d\Omega^{4\pi} + \left(\Omega^{4\pi}/4\pi\right)d\omega$$

$$= -SdT - p^\alpha \omega R_0^2 dR_0 - Nd\mu^\alpha + \left(\Omega/\omega\right)d\omega. \tag{23}$$

By comparing Equations (21) and (23), we obtain

$$\sigma = \Omega/\omega \equiv \left(E - TS - \mu^\alpha N\right)\big/\omega. \tag{24}$$

In previous papers [2,3,5,9,11,13], the last expression was given, despite that the mid expression is conceptually meaningful. This equation is the equation obtained from the relation on the basis of the fact that when the solid angle is multiplied by λ, the grand potential

$\Omega\left(T, R_0, \mu^\alpha; \omega\right)$ is transformed as

$$\Omega\left(T, R_0, \mu\alpha; \lambda\omega\right) = \lambda\Omega\left(T, R_0, \mu\alpha; \omega\right)$$

(Euler's theorem). We note that Nishioka [11] derived the same equation by applying Euler's theorem to E.

Interfacial Tension

In Equation (21), existence of $\sigma d\omega$ is due to the nucleus. Therefore, one can write

$$\sigma d\omega = -p^\alpha dV^\alpha - p^\beta dV^\beta + \gamma dA + CdR, \tag{25}$$

(pay attention on the independent variables). The first two terms are of the hypothetical system defined in Section 3.1. The last two terms are for forming interfacial structure after the formation of the hypothetical system. As mentioned above, we note that a term depending on the derivative of the curvature radius, dR, appears. This term, also as mentioned above, vanishes if the surface of tension is taken as the dividing surface.

Let us go forward the argument by taking the surface of tension as the dividing surface. Using the equation obtained by putting C=0 in Equation (25), we rewrite Equation (21) into

$$d\Omega = -SdT - p^{\alpha}\omega R_0^2 dR_0 - Nd\mu$$
$$-p^{\alpha}dV^{\alpha} - p^{\beta}dV^{\beta} + \gamma dA. \tag{26}$$

The fundamental equation for the hypothetical system is just the addition of both of Equation (17):

$$d\left(\Omega^{\alpha} + \Omega^{\beta}\right) = -\left(S^{\alpha} + S^{\beta}\right)dT - p^{\alpha}dV^{\alpha}$$
$$-p^{\beta}dV^{\beta} - \left(N^{\alpha} + N^{\beta}\right)d\mu. \tag{27}$$

Subtracting Equation (27) from Equation (26), we have the fundamental equation for the superficial grand potential $\Omega^s = \Omega - \left(\Omega^{\alpha} + \Omega^{\beta}\right)$:

$$d\Omega^s = -S^s dT - N^s d\mu^{\alpha} + \gamma dA$$
$$= -S^s dT - N^s d\mu^{\alpha} + \gamma R^2 d\omega, \tag{28}$$

where $S^s = S - \left(S^{\alpha} + S^{\beta}\right)$ and $N^s = N - \left(N^{\alpha} + N^{\beta}\right)$ are, respectively, the superficial entropy and the superficial number of molecules. In this equation $-p^{\alpha}\omega R_0^2 dR_0$ has been eliminated because the state of the interface is independent of the selection of R_0; in other words, R_0 has been fixed at the position $R_0 \gg R$. Euler's relation obtained from the fact that Ω^s is transformed as $\Omega\left(T, \mu^{\alpha}; \lambda\omega\right) = \lambda\Omega\left(T, \mu^{\alpha}; \omega\right)$ when ω is multiplied by λ as $\omega \to \lambda\omega$ is

$$\gamma A = \Omega^s = E^s - TS^s - N^s \mu. \tag{29}$$

To derive this equation, one can use the same method to derive Equation (24). From Equation (29), the interfacial tension γ is revealed to be the superficial grand potential per unit area of the interface. Introducing the superficial quantities per unit area of the interface, $e^s = E^s / A, S^s = S^s / A$, and $\Gamma = N^s / A$, we have

$$\gamma = \Omega^s / A = e^s - Ts^s - \Gamma\mu. \tag{30}$$

The last expressions in Equations (29) and (30) have already be given in previous papers [3,4,5,9,11-13,15,20, 21,31]. In those papers, except for [12,20,21,31]—Rusanov et al. [20] explicitly stated, however, the word of the superficial grand potential did not appear.

Gibbs-Duhem Relation for Interface

A general way to obtain the Gibbs-Duhem relation is to take differential of Euler's relation and subtract the fundamental equation. For the interface, the same procedure is possible; we can have the Gibbs-Duhem relation for the interface

$$d\gamma = -s^s dT - \Gamma d\mu^\alpha, \tag{31}$$

by taking differential of Equation (29) and subtract the first equation of Equation (28) and dividing by A. We can, also, obtain Equation (31) by direct differentiation of Equation (30) and using the fundamental equation for e^s. From Equation (31) we have Equation (15) or

$d\gamma = -\Gamma d\mu \left(T = \text{const.}\right)$. This is the Gibbs adsorption isotherm.

SUMMARY

We have given a grand potential formalism for the interfacial thermodynamics. It is revealed that the work of formation of a critical nucleus is equal to the grand potential difference. This makes a point of view clearer overwhelmingly than regarding the work of formation of the nucleus as the Gibbs energy difference. Also, the interfacial tension is revealed to be defined as the superficial grand potential per unit area of the interface. Although equivalent form was given previously [3-5,9,11, 13], this paper has explicitly closed up the grand potential property for the first time.

ACKNOWLEDGEMENTS

This paper is base on a lecture [in Japanese] at the 35th research meeting on the crystal growth (Toronkai) held by Japanese Association for Crystal Growth on Sept. 7-9, 2011 at Shimotsuma, Japan.

REFERENCES

1.   Gibbs, J.W. (1993) The scientific papers of J. Willard Gibbs, thermodynamics. Ox Bow, Woodbridge.

2.   Tolman, R.C. (1948) Consideration of the Gibbs theory of surface tension. The Journal of Chemical Physics, 16, 758-774. doi:10.1063/1.1746994

3.   Hill, T.L. (1951) On Gibbs theory of surface tension. The Journal of Chemical Physics, 19, 1203-1203. doi:10.1063/1.1748502

4.   Buff, F.P. (1951) The spherical interface. I. Thermodynamics. The Journal of Chemical Physics, 19, 1591-1594. doi:10.1063/1.1748127

5.   Hill, T.L. (1952) Statistical thermodynamics of the transition region between two phases. I. Thermodynamics and quasi-thermodynamics. Journal of Physical Chemistry, 56, 525-531. doi:10.1021/j150496a027

6.   Kondo, S. (1955) A statistical-mechanical thory of surface tension of curved surface layer I. Journal of the Physical Society of Japan, 10, 381-386. doi:10.1143/JPSJ.10.381

7.   Kondo, S. (1956) Thermodynamical fundamental equation for spherical interface. The Journal of Chemical Physics, 25, 662-669. doi:10.1063/1.1743024

8. Plesner, I.W. (1964) Statistical thermodynamics of spherical droplets. The Journal of Chemical Physics, 40, 1510- 1517. doi:10.1063/1.1725355

9. Nishioka, K. (1977) Thermodynamics of a liquid microcluster. Physical Review A, 16, 2143-2152. doi:10.1103/PhysRevA.16.2143

10. Wilemski, G. (1984) Composition of the critical nucleus in multicomponent vapor nucleation. Journal of Chemical Physics, 80, 1370-1372. doi:10.1063/1.446822

11. Nishioka, K. (1987) Thermodynamics formalism for a liquid microcluster in vapor. Physical Review A, 36, 4845- 4851. doi:10.1103/PhysRevA.36.4845

12. Voorhees, P.W. and Johnson, W.C. (1989) The thermodynamics of a coherent interface. The Journal of Chemical Physics, 90, 2793-2801. doi:10.1063/1.455928

13. Nishioka, K. and Kusaka, I. (1992) Thermodynamics formulas of liquid phase nucleation form vapor in multicomponent systems. The Journal of Chemical Physics, 96, 5370-5376. doi:10.1063/1.462721

14. Oxtoby, D.W. and Kashchiev, D. (1994) A general relation between the nucleation work and the size of the nucleus in multicomponent nucleation. The Journal of Chemical Physics, 100, 7665-7671. doi:10.1063/1.466859

15. Debeneditti, P.G. and Reiss, H. (1998) Reversible work of formation of an embryo of a new phase within a uniform macroscopic mother phase. The Journal of Chemical Physics, 108, 5498-5505. doi:10.1063/1.475938

16. Laaksonen, A., McGraw, R. and Vehkamäki, H. (1999) Liquid-drop formalism and free-energy surface in binary homogeneous nucleation theory. The Journal of Chemical Physics, 111, 2019-2027. doi:10.1063/1.479470

17. Kashchiev, D. (2003) Thermodynamically consistent description of the work to form a nucleus of any size. The Journal of Chemical Physics, 118, 1837-1851. doi:10.1063/1.1531614

18.   Schmelzer, J.W.P., Baidakov, V.G. and Boltachev, G.S. (2003) Kinetics of boiling in binary liquid-gas solutions: Comparison of different approaches. The Journal of Chemical Physics, 119, 6166-6183. doi:10.1063/1.1602066

19.   Kashchiev, D. (2004) Multicomponent nucleation: Thermodynamically consistent description of the nucleation work. The Journal of Chemical Physics, 120, 3749-3758. doi:10.1063/1.1643711

20.   Rusanov, A.I., Shchekin, A.K. and Tatyanenko, D.V. (2009) Grand potential in thermodynamics of solid bodies and surfaces. The Journal of Chemical Physics, 131, 161104. doi:10.1063/1.3254324

21.   Corti, D.S., Kerr, K.J. and Torabi, K. (2011) On the interfacial thermodynamics of nanoscale droplets and bubbles. The Journal of Chemical Physics, 135, 024701. doi:10.1063/1.3609274

22.   Tolman, R.C. (1949) The effect of droplet size on surface tension. The Journal of Chemical Physics, 17, 333-337. doi:10.1063/1.1747247

23.   Koenig, F.O. (1950) On the thermodynamic relation between surface tension and curvature. The Journal of Chemical Physics, 18, 449-459. doi:10.1063/1.1747660

24.   Buff, F.P. and Kirkood, J.G. (1950) Remarks on the surface tension of small droplets. The Journal of Chemical Physics, 18, 991-992. doi:10.1063/1.1747829

25.   Buff, F.P. (1955) Spherical interface. II. Molecular theory. The Journal of Chemical Physics, 23, 419-427. doi:10.1063/1.1742005

26.   Bogdan, A. (1997) Thermodynamics of the curvature effect on ice surface tension and nucleation theory. The Journal of Chemical Physics, 106, 1921-1929. doi:10.1063/1.473329

27.   McGraw, R. and Laaksonen, A. (1997) Interfacial curvature free energy, the Kelvin relation, and vapor-liquid nucleation rate. The Journal of Chemical Physics, 106, 5284-5287. doi:10.1063/1.473527

28. Koga, K., Zeng, X.C. and Shchekin, A.K. (1998) Validity of Tolman's equation: How large should a droplet be? The Journal of Chemical Physics, 109, 4063-4070.doi:10.1063/1.477006

29. Bartell, L.S. (2001) Tolman's δ, surface curvature, compressibility effects, and free energy of drops. Journal of Physical Chemistry B, 105, 11615-11618. doi:10.1021/jp011028f

30. Blokhuis, E.M. and Kuipers, J. (2006) Thermodynamic expressions for the Tolman length. The Journal of Chemical Physics, 124, 074701. doi:10.1063/1.2167642

31. Tröster, A., Ottel, M., Block, B., Virnau, J.P. and Binder, K. (2012) Numerical approaches to determine the inter facial tension of curved interface from free energy calculations. The Journal of Chemical Physics, 136, 064709. doi:10.1063/1.3685221

32. Nishioka, K., Tomino, H., Kusaka, I. and Takai, T. (1989) Curvature dependence of the interfacial tension in binary nucleation. Physical Review A, 39, 772-782.doi:10.1103/PhysRevA.39.772

33. Yang, A.J.-M. (1983) Free energy for the heterogeneous systems with spherical interfaces. The Journal of Chemical Physics, 79, 6289-6293. doi:10.1063/1.445734

34. Mutaftschiev, B. (1993) Nucleation theory. In: Hurle, D.T.J., Ed., Handbook of Crystal Growth, Part 1a, Chapter 4, Elsevier, Amsterdam.

35. Saito, Y. (1996) Statistical physics of crystal growth. World Scientific, Singapore.

36. Markov, I.V. (2003) Crystal growth for beginners: Fundamentals of nucleation, crystal growth and epitaxy. 2nd Edition, World Scientific, Singapore.doi:10.1142/9789812796899

37. Doyle, G.J. (1961) Self-nucleation in the sulfuric acidwater system. The Journal of Chemical Physics, 35, 795- 799. doi:10.1063/1.1701218

38. Kashchiev, D. (1982) On the relation between nucleation work, nucleus size, and nucleation rate. The Journal of Chemical Physics, 76, 5098-5102. doi:10.1063/1.442808

39. Laaksonen, A., Kulmala, M. and Wanger, P.E. (1993) On the cluster compositions in the classical binary nucleation theory. The Journal of Chemical Physics, 99, 6832- 6835. doi:10.1063/1.465827

40. Vilsanen, Y. and Strey, R. (1994) Homogeneous nucleation rates for n-butanol. The Journal of Chemical Physics, 101, 7835-7843. doi:10.1063/1.468208

41. Schmelzer, J.W.P., Schmelzer Jr., J. and Gutzow, I.S. (2000) Reconciling Gibbs and van der Waals: A new approach to nucleation theory. The Journal of Chemical Physics, 112, 3820-3831. doi:10.1063/1.481595

42. Bowels, R.K., Reguera, D., Djikaev, Y. and Reiss, H. (2001) A theorem for inhomogeneous systems: The generalization of the nucleation theorem. The Journal of Chemical Physics, 115, 1853-1866.

43. Auer, S. and Frenkel, D. (2001) Prediction of absolute crystal-nucleation rate in hard-sphere colloids. Nature, 409, 1020-1023. doi:10.1038/35059035

44. Gasser, S., Weeks, E.R., Schofield, A., Pusey, P.N. and Weitz, D.A. (2001) Real-space imaging of nucleation and growth in colloidal crystallization. Science, 292, 258-262. doi:10.1126/science.1058457

45. Kawasaki, T. and Tanaka, H. (2010) Formation of a crystal nucleation from liquid. Proceedings of the National Academy of Sciences of the United States of America, 107, 14036-14041.

46. Nishioka, K., Mori, A., Takano, K.J., Kaishita, Y. and Narimatsu, S. (1999) Pressure-dependence of the interfacial tension of a critical nucleus in the binary-ideal solution. Journal of Crystal Growth, 200, 592-598. doi:10.1016/S0022-0248(98)01391-8

47. Callen, H.B. (1985) Thermodynamics and an introduction to thermostatistics. 2nd Edition, Wiley, New York.

48. Landau, L.D. and Lifshitz, E.M. (1989) Statistical physics. 3rd Edition, Pt. 1, Pergamon, Oxford.

49. Yang, A.J.-M. (1985) The thermodynamical stability of the heterogeneous system with a spherical interface. The Journal of Chemical Physics, 82, 2082-2085.doi:10.1063/1.448344

50. Barrett, J. (1999) First-order correction to classical nucleation theory: A density functional approach. The Journal of Chemical Physics, 111, 5938-5946. doi:10.1063/1.479889

51. Hurbý, J., Labetski, D.G. and van Dongen, M.E.H. (2007) Gradient theory computation of the radius-dependent surface tension and nucleation rate for n-nonane clusters. The Journal of Chemical Physics, 127, 164720. doi:10.1063/1.2799515

52. Mori, A. and Suzuki, Y. (2013) Vanishing linear term in chemical potential difference in volume term of work of critical nucleus formation for phase transition without volume change. Journal of Crystal Growth, 375, 16-19. doi:10.1016/j.jcrysgro.2013.04.005

Turbomachinery for the Air Management and Energy Recovery in Fuel Cell Gas Turbine Hybrid Systems

A. Traverso, L. Magistri, and A.F. Massardo

DIMSET (TPG), University of Genoa, Via Montallegro 1, 16145, Genoa, Italy

ABSTRACT

High temperature fuel cells (MCFCs and SOFCs) can operate at atmospheric or pressurised conditions. In both cases, system performance can be significantly improved when the fuel cells are integrated with proper devices, which are designed to provide the necessary air inlet conditions and to recover the exhaust gas energy. This paper presents a review of modelling and design issues for the integration of turbomachinery with the fuel cell system, because turbomachinery is the most promising technology for coping with the high temperature fuel cell requirements. Since the gas turbine expander performance is significantly influenced by exhausts compositions, analytical approach is undertaken for properly modelling composition influence on expander performance, and results are presented to demonstrate the quantitative influence of the system parameters on the performance. The analysis covers the three main aspects of performance evaluation: the on-design, the off-design and, as a final mention, the control of the fuel cell hybrid systems.

INTRODUCTION

Fuel cells are electrochemical reactors that allow an efficient and ecological conversion of energy. High efficiency close to 50% also at part-load conditions, and low pollution, make fuel cells a very interesting system for distributed power generation. In particular, high temperature fuel cells such as Solid Oxide Fuel Cells (SOFC) are analysed in this paper. The air management, the pressurisation and the energy recovery from such fuel cells typically need the integration of the fuel cell stack with external devices, such as turbomachinery. The integration of existing or specifically designed turbomachines into the hybrid system is expected to:

- recover energy from the fuel cell exhaust gases, since they are at high temperature and, for pressurised stack, at high pressure;
- to pressurise the stack itself;
- to generate extra electrical power (up to 25% of the fuel cell power) to increase the system performance from the point of view of the efficiency, the specific power and the costs;

Several options can be investigated for supplying the fuel cell system with the flows at the proper conditions. The possibilities analysed in the present paper refer to electrically driven compressors for anode (fuel) and cathode side (air), turbocharger, simple cycle gas turbine, regenerated gas turbine (micro gas turbine). Depending on the size, turbomachinery can be of either axial or radial type. The present interest in sub-MW fuel cell packages led the authors to mainly stress the radial type turbomachinery: Fig. 1 illustrates the conceptual layout of a radial compressor and turbine mounted on single rotating shaft. Notwithstanding, the conclusions and the results presented can be well extended also to larger size plants employing axial turbomachinery. The following paragraphs present an overview on existing radial turbomachinery and their typical applications: afterwards, a detailed analysis of expander performance representation taken into account the fuel cell exhaust composition is undertaken, this being a relevant issue in predicting performance of fuel cell gas turbine hybrid system.

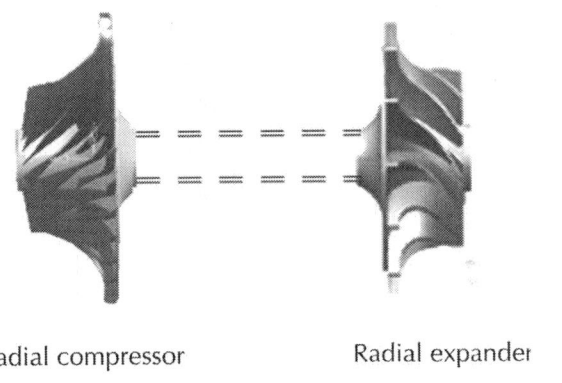

Radial compressor   Radial expander

Figure 1: Schematics of a single-shaft radial compressor and expander.

Existing Turbomachinery

Turbochargers are very commonly employed in both small and large piston engines for enhancing their performance and emissions. Fig. 2 illustrates an example of a typical car engine turbocharger, with the fresh air compressor on the left, and the hot exhausts expander on the right. Fig. 3 shows the conceptual layout of turbocharger integration with an internal combustion engine, where an intercooler might be applied between compressor discharge and piston engine air inlet. The air is usually pressurised by a single-stage radial compressor, while the turbine is either constituted of a single-stage radial or axial expander, depending on the size. Compressor and expander are usually mounted on a shaft as close as possible, making the heat exchange between them non-negligible. Turbochargers present interesting features for integration with hybrid MCFC systems, where the operational temperature of the fuel cell system is similar to the cylinder outlet temperature; on the other hand, they are unsuitable for hybrid SOFC systems, where the fuel cell temperature is significantly higher, and its use would require a redesign of the expander. Moreover, turbochargers are not equipped with a motor/generator, this being an important device for performance enhancement, start-up and control of the hybrid system. In this respect, microturbines are a more apt technology, even if the integration of existing gensets with the hybrid system requires the solving of several problems. Turbocharger delivery pressure and air flow rate can vary from 1.5 to 2.5 bar (absolute) and 0.1–1.0 kg/s for automotive applications to 3–5 bar (absolute) and 1.0–40.0 kg/s for large navy applications.

Figure 2: Picture of an actual turbocharger for piston engines (courtesy of Garrett).

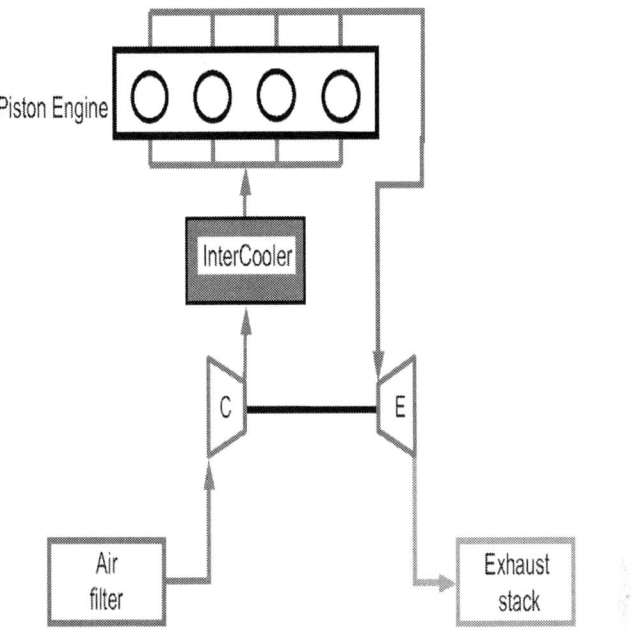

Figure 3: Schematics of a piston engine equipped with an intercooled turbocharger.

In power engineering research, microturbines are very attractive mainly for two possible applications:

- experimenting on and testing complex gas turbine-based power plants scaled-down to a <1 MW configuration
- integration with high temperature fuel cells (typically MCFCs and SOFCs)

The second application is, of course, still at an R&D stage, because no commercial hybrid systems based on high temperature fuel cells are available yet.

However, microturbines normally rotate at very high speeds: typically, permanent magnetic electrical generators are directly mounted on the rotating shaft, generating power at frequencies significantly higher than the grid. Thus, power electronics is often required for grid connection: Fig. 4 illustrates the conceptual power chain from the shaft to the grid. The typical total efficiencies of such a conversion are in the order of 80–90%, significantly affecting the performance of the machine. While the generator normally performs well (about 97–98% electrical efficiency), the power electronics unit can present an efficiency of about 85–90%, which, in turn, reduces the overall efficiency. Power electronics frees the power output frequency from the shaft rotational speed, because it is determined by an embedded microprocessor.

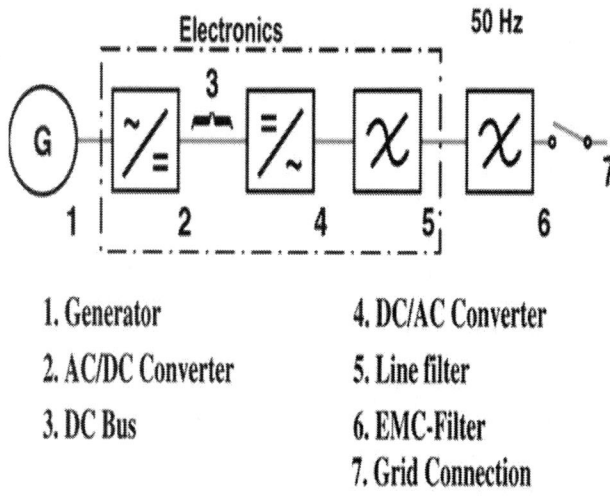

1. Generator
2. AC/DC Converter
3. DC Bus
4. DC/AC Converter
5. Line filter
6. EMC-Filter
7. Grid Connection

Figure 4: Schematics of a typical electronic power unit.

Conventional gas turbines typically operates in simple Brayton cycle. On the other hand, microturbines are typically arranged in recuperated cycle, as shown in Fig. 5, in order to increase the electrical efficiency. In such a configuration, the microturbine is coupled with the recuperator [1], which transfers heat from the exhausts to the compressed air, so reducing (usually about halving) the amount of fuel required to achieve the nominal TIT before the turbine nozzle (when the turbine has a radial design, the nozzle may also be unvaned).

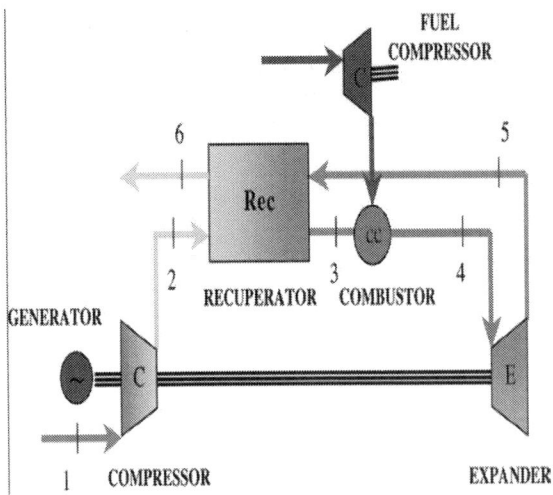

Figure 5: Schematics of a recuperated microturbine: 1) intake air, 2) compressed air, 3) heated compressed air delivered to the combustor, 4) combustor outlet exhausts, 5) expander exhausts, 6) cycle exhausts.

TURBOMACHINERY FOR HYBRID SYSTEMS

Radial Compressor

Radial compressors represent an established technology for gas turbine cycles of small and medium size, having well-known applications both in piston engine (e.g.: turbocharger) and aircraft applications (e.g.: helicopter engine, APU). The choice focuses on radial compressors because it is expected that they will be employed by the first generation of commercial fuel cell hybrid systems, which are likely to address the 1 MW size applications for stationary power production [2]. In the middle- and long-term, also axial compressors have the potential for integration with fuel cells, but the plant size will need to be higher.

Typical maps of radial compressors are reported in Fig. 6 and Fig. 7: the former shows pressure ratio against corrected mass flow for different corrected rotational speeds; the latter illustrates the trends of compressor adiabatic efficiency against pressure ratio for different corrected rotational speeds.

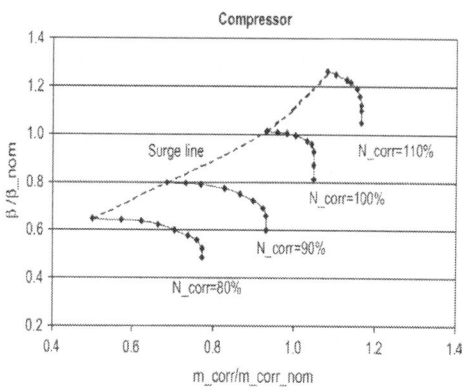

Figure 6: Typical non-dimensional performance map: radial compressor pressure ratio versus corrected mass flow.

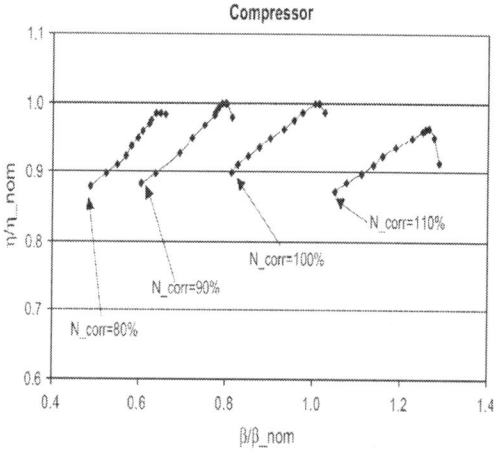

Figure 7: Typical non-dimensional performance map: radial compressor isentropic efficiency versus pressure ratio.

Within a hybrid fuel cell gas turbine system, the compressor has mainly to provide air to the stack and to pressurise the fuel cell system. The former is necessary for both atmospheric and pressurised hybrid system configurations, while the latter is concerned only with the pressurised layout, where the entire fuel cell system is contained in a pressure vessel. The main issue in dealing with the compressor is the possible situation where the operating point on the compressor map moves beyond the stability limit, and surge occurs. In an actual plant, this is a very dangerous condition, which must be avoided at all because it can cause unexpected fluctuations of pressure, mass flow and, finally, shaft speed that will rapidly destroy the machine. Apart from turbomachinery damage, another major source of concern is represented by the fuel cell itself, which can undergo fast cycling variations in pressure, causing possible damage to the stack, where the pressure difference between the anode and cathode sides should be kept as low as possible, in order to minimise the mechanical stress to the ceramic parts. So, compressor surge should always be avoided in actual hybrid systems, and proper safety measures should be implemented in the plant (e.g.: blow-off valve) in order to prevent it happening, by keeping the surge margin within safe limits. The surge margin, Kp, is defined by Eq. (1). Commonly, Kp should always be kept above 5–10% (Kp > 1.05–1.10) to have a safe stable impeller operation.

$$Kp = \frac{\beta_{surge}/\dot{m}_{surge}}{\beta/\dot{m}} > 1.05 \div 1.10 \tag{1}$$

Eq. (1) is a standard formulation for compressor surge margin (pressure ratio over mass flow at surge against the same ratio at the present operating conditions), while others can include additional parameters, such as the impeller outlet temperature.

A final note is required on the compressor heat exchange. Two aspects need to be considered:

- the heat exchange during steady-state operation;
- the heat exchange during transients.

The former is due to the proximity of the compressor to the expander, allowing significant heat exchange to take place between the hot exhausts and the cold compressed air: a 6–10 °C increase in temperature at the compressor outlet is possible (and a 4–8 °C decrease in temperature at the turbine outlet). The latter becomes relevant when strong temperature variations are applied to the working fluid, so that the impeller (compressor in Fig. 1) behaves as a perfectly-finned heat exchanger where the fins are constituted by the blades. This causes a non-negligible transient heat exchange with the rotor skin, until new steady-state conditions are reached. Such transient operating conditions are normally verified during the start-up and shut-down procedures of gas turbines (and microturbines as well). As a result, the error in the estimation of the rotational speed could be up to 25% [3], because the model tends to show prompter responses. In a hybrid system, the delay in the air delivery temperature introduced by the compressor and the air ducts can be helpful for smoothing out the thermal gradients.

Radial Turbine (Expander)

Radial turbines are usually coupled with radial compressors, to achieve a compact one-stage to one-stage machine. For the performance assessment, the turbine characteristic maps are usually provided for a reference working fluid (e.g.: dry air): in this case, corrections based on the similitude theory should be applied when the flow composition is changed. Fig. 8 and Fig. 9 shows typical performance maps for a radial turbine: the former reports pressure ratio against corrected mass flow for different corrected rotational speeds (rotational speed influence is minor); the latter illustrates the trends of expander adiabatic efficiency against pressure ratio for different corrected rotational speeds.

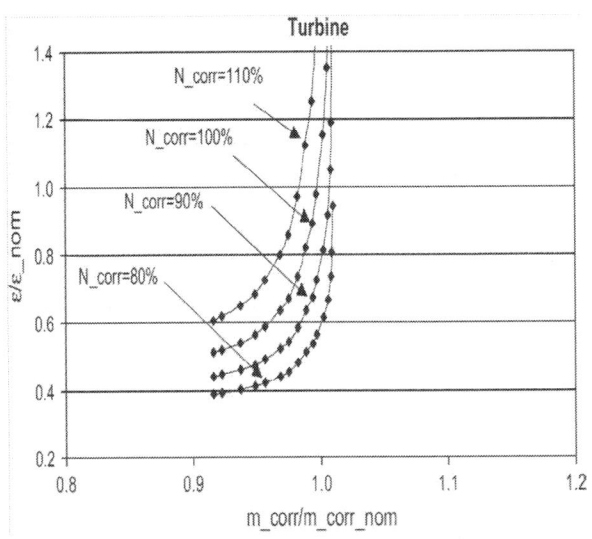

Figure 8: Typical non-dimensional performance map: radial turbine expansion ratio versus corrected mass flow.

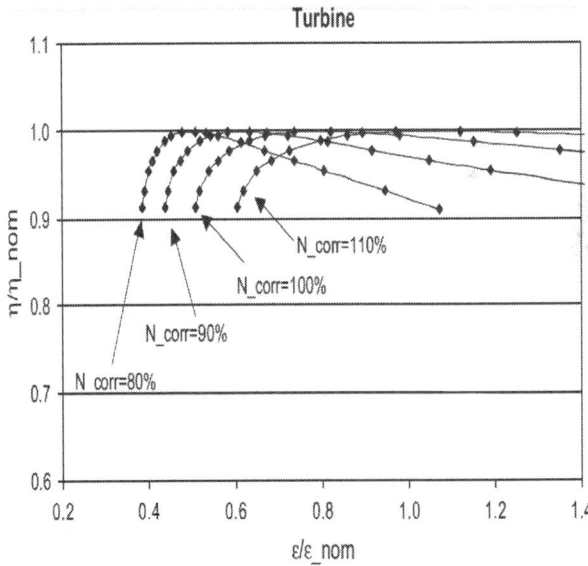

Figure 9: Typical non-dimensional performance map: radial turbine isentropic efficiency versus expansion ratio.

For the study of turbine time-dependent behaviour, differently from that of the compressor, where instability is a major issue not only for the component itself but for the entire cycle, dynamic analyses can be useful for two purposes: the study of the interaction of nozzle/rotor, and the study of the interaction of a pulsating flow with the expander (turbocharger case). In both cases the 1-D approach for the expander can only provide partial information, because such phenomena investigated for performance optimisation involve complex 2-D and 3-D effects, which cannot be properly taken into account in a 1-D approach. Nevertheless, under normal operating conditions, a 1-D approach or, even, a 0-D approach, based on characteristic maps, can be satisfactory for cycle simulations.

Similitude Theory

Similitude theory needs to be applied to compressor and expander models whenever the boundary conditions (e.g.: inlet temperature and pressure) and fluid properties (e.g.: gas composition) change with respect to the conditions used for measuring the characteristic curves (typically, dry air at fixed inlet temperature and pressure). The composition variation of expander exhausts is particular relevant for fuel cell hybrids: a new comprehensive theory is thus presented in the following.

The theory was developed for radial machinery, but, in principle, could be extended to axial turbines as well. Valuable background for the basis of the analysis presented here can be found in [4], while the complete theory is presented in [3].

Once the performance curves are known for a given working fluid and boundary conditions, similitude considerations become necessary when the turbomachinery is operated under different operating conditions, such as different inlet conditions or working fluid chemical composition.

The working fluid is regarded as a perfect gas, with constant c_p and c_v. The fluid is considered to follow a real adiabatic transformation.

The non-dimensional performance parameters are expressed functionally as:

$$\frac{p_{0in}}{p_{0out}}, \eta, \frac{\Delta T_0}{T_{0in}} = F\left(\frac{\dot{m}\sqrt{RT_{0in}}}{D^2 P_{0in}}, \frac{ND}{\sqrt{KRT_{0in}}}, Re, k\right)$$

(2)

Commonly, at this point, R and k are dropped from Eq. (2) because they are considered to be constant[4] and [5]. This is not accurate in the hybrid system case because the aim here is to consider significant variations in the chemical composition of the working fluid. Instead, it is possible to delete two other parameters: D, because the machine is considered to be fixed with no alterations in the geometrical design, and Re, because turbomachinery usually operates at high Reynolds numbers, making its influence negligible. So, the final functional relationship becomes:

$$\varepsilon, \eta, \frac{\Delta T_0}{T_{0in}} = F\left(\frac{\dot{m}\sqrt{RT_{0in}}}{P_{0in}}, \frac{N}{\sqrt{KRT_{0in}}}, k\right)$$

(3)

Eq. (3) clearly expresses the influence of k and R. The problem in implementing such a relationship is that k has a separate effect, usually unknown. To overcome this limitation, the following approach, based on the superposition of the effects, is proposed. The aim is double:

- first of all, to define two-parameters, to be used as arguments in Eq. (3), which could well represent the effect of k inside the parameter itself,

- then, to apply the k influence separately to the resultant parameters (left side of Eq. (3)).

In this way, it is possible to use a more effective two-parameter relationship with a k-correction to account for the composition effects.

The rotational speed parameter already accounts for k variations by including the speed of sound in the denominator, while the flow coefficient needs to be revised. Considering the mass flow relationship[4] and [5], when applied to the expander inlet, we have:

$$\frac{\dot{m}\sqrt{RT_{0in}}}{p_{0in}} = \frac{Ma\sqrt{k}}{f^\chi} = \dot{m}_{corr}$$

(4)

where

$$[\chi = k + 1/2(k-1)], f = (1 + \tfrac{k-1}{2}Ma^2)$$

A straightforward assumption is to consider the nozzle of the expander as always choked (Ma = 1), neglecting the losses in total pressure. In this way Eq. (4) reduces to

$$\frac{\dot{m}\sqrt{RT_{0in}}}{p_{0in}} = \frac{\sqrt{k}}{\left(\frac{k+1}{2}\right)^\chi} = \dot{m}_{corr\_crit}$$

(5)

It follows that, whatever the value of k, we have:

$$\frac{\dot{m}\sqrt{RT_{0in}}}{p_{0in}} \cdot \frac{1}{\dot{m}_{corr\_crit}} = constant$$

(6)

So, this flow parameter well represents the influence of k.

This approach is likely to become less accurate as choking conditions are no longer verified in the nozzle. However, it can be demonstrated numerically that the ratio $\dot{m}_{corr}/\dot{m}_{corr\_crit}$, having fixed Mach number to whatever value in the [0,1] range, varies little with k. This demonstrates that Eq. (6) is quite accurate in taking into account the "k" effect (see Fig. 10 and Table 1).

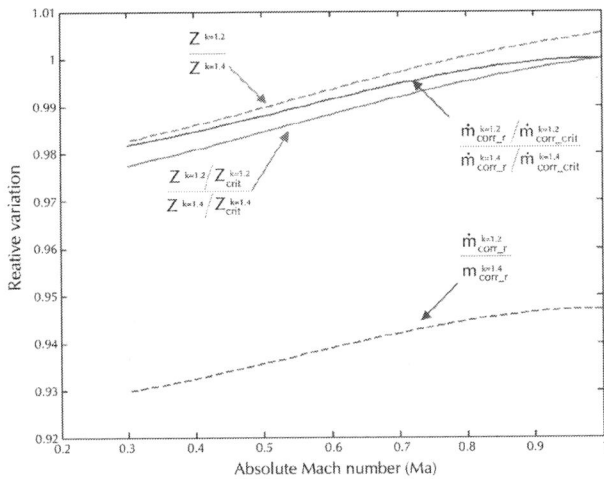

Figure 10: Variation of non-dimensional parameters after a variation of k of 15% (from 1.4 to 1.2).

Table 1: Main assumptions and conclusions of Fig. 10

Description	Value	Comment
Hypothesis		
Ma_r	1.0	Rotor is chocked
Ma_u	1.0	
$\alpha_{vol\_out\_crit}$	60°	at k = 1.4
k	1.4 ⇒ 1.2	

Description	Variation at Mach = 1	Comment
Results		
k	−15%	Driving parameter
$\dot{m}_{corr-r}$	−5.3%	Primary effect
Z	+0.5%	Secondary effect
$\alpha_{vol\ out\ crit}$	−0.8%	60° ⇒ 59.5°

If the choking conditions are not verified in the nozzle, two alternatives are possible: the choked point is placed in the rotor (common case for free volute radial turbines – no stator vanes), the turbine is not choked (low flow regime). Since, in the latter case, the turbine behaviour that has to be represented is more complex, the approach is developed for a choked turbine and then extended to the non-choked conditions. The passages to obtain the final expressions can be found in [3], and they demonstrate that Eq. (6) is reasonably valid also for the case in which choking conditions are verified at the beginning of the rotor.Fig. 11 shows the naming convention used for the equations.

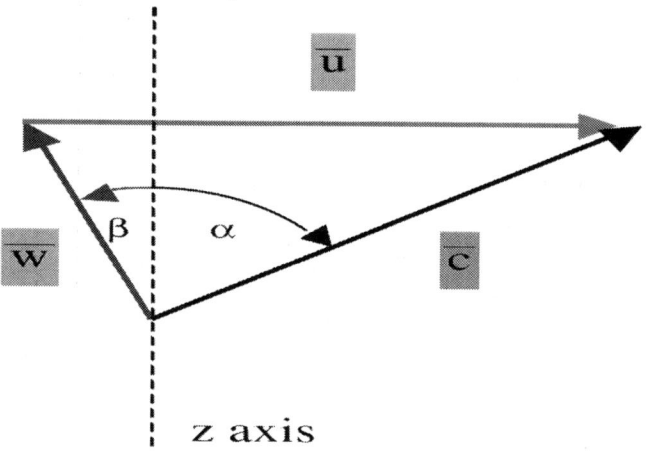

Figure 11: Naming convention.

Let us make three hypotheses:

- the turbine rotor is choked very close to the flow inlet;
- the work exchanged by the flow from the rotor inlet to the choked point is negligible;

the losses (especially in total pressure) from the rotor inlet to the choked point are negligible.

The last assumption is particularly acceptable close to the design point, while it becomes less realistic under off-design conditions.

It is possible to write the mass flow equation in the rotor using relative terms [6]:

$$\frac{\dot{m}\sqrt{RT_{0r}}}{A^* \cos \beta p_{0r} f_{ur}^{\chi}} = \frac{Ma_r \sqrt{k}}{f_r^{\chi}} \tag{7}$$

where A^* is the flow section perpendicular to the radial direction, and b the relative flow angle (seeFig. 11), assumed to be equal to the constructive angle.

Then, defining M as reported by (8), the following expression can be obtained:

$$M = \frac{Ma_u Ma}{\sqrt{f}} \cdot \sin \alpha \cdot (k-1) \quad \text{where} \quad Ma_u = \frac{u}{\sqrt{kRT_0}} \tag{8}$$

$$\frac{\dot{m}\sqrt{RT_{0in}}}{A^* \cos \beta p_{0in}} \cdot \frac{1}{\dot{m}_{corr\_crit}} = [(1-M)f_{ur}]^{\chi} \frac{Ma_r \sqrt{k}}{f_r^{\chi} \cdot \dot{m}_{corr\_crit}} \tag{9}$$

At rotor choking conditions ($Ma_r = 1$), Eq. (9) differs from Eq. (6) due to the presence of the term in square brackets, which accounts for the rotational speed. This is the only difference and so it is enough to study the influence of k over such a term, called Z.

$$Z = [(1-M)f_{ur}]^{\chi} = \left(1 - \frac{Ma_u Ma}{\sqrt{f}} \cdot \sin \alpha \cdot (k-1) + \frac{k-1}{2} Ma_u^2\right)^{\chi} \tag{10}$$

where

$$Ma_u = \frac{u}{\sqrt{kRT_{0in}}}.$$

Since the blockage in the rotor is likely to happen when a vaneless nozzle is employed, the absolute angle α_{vol_out}, at the volute outlet (rotor inlet), can vary with k, because it is not fixed by geometrical constraints. The angle can be determined by applying the free-vortex equation to the turbine volute, as a result of radial equilibrium:

$$C_v r = C_{vin} r_{in} = cons\,tan\,t \tag{11}$$

where c_ϑ is the tangential velocity in the volute ($= c \sin \alpha$).

Let us consider equal the static and total properties at the volute inlet. By applying the continuity equation and defining α as in Eq. (12), where c_r is the radial velocity in the volute ($= c \cos \alpha$), it is possible to rewrite Eq. (10) into Eq. (13) introducing the new term Z_{crit} to the denominator, which represents the numerator calculated for Ma $= 1$ and Ma$_u = 1$.

$$\frac{c_\vartheta}{c_r} = tg\alpha$$

(12)

$$\Rightarrow \frac{Z}{Z_{crit}} = \frac{\left(1 - \frac{Ma_u Ma}{\sqrt{f}} \cdot \sin \alpha_{vol\_out} \cdot (k-1) + \frac{k-1}{2} Ma_u^2 \right)^\chi}{\left(1 - \frac{1 \cdot 1}{(\frac{k+1}{2})^{\frac{1}{2}}} \cdot \sin \alpha_{vol\_out\_crit} \cdot (k-1) + \frac{k-1}{2} \cdot 1 \right)^\chi}$$

(13)

Hence it is possible to express Eq. (9) as Eq. (14).

$$\frac{\dot{m}\sqrt{RT_{0in}}}{A^* \cos\beta \; p_{0in}} \cdot \frac{1}{\dot{m}_{corr\_crit}} \cdot \frac{1}{Z_{crit}} = \frac{z}{z_{crit}} \cdot \frac{\dot{m}_{corr\_r}}{\dot{m}_{corr\_crit}}$$

(14)

The question is whether the Z term is necessary to take the k variations into account. For assessing this at the choked point (Ma$_r$ $= 1$), let's fix, for this analysis, $k = 1.4$, Ma$_u = 1$ and $_{vol\_out\_crit} = 60°$. The values of Z and $\dot{m}_{corr\_r}$ are then calculated for $k = 1.4$ and $k = 1.2$, for different Mach numbers (Ma). The results, making Ma vary in the range [0.3; 1], are reported in Fig. 10 and Table 1. It should be observed that, within the similitude theory, keeping Ma$_u$ constant is conceptually rigorous, because it involves a constant value in the second parameter of (3), which involves that the two cases are "similar".

The numerical results demonstrate that for properly considering the effects of k variations:

- the Z term can be neglected;

- the introduction of $\dot{m}_{corr\_crit}$ well represents the influence of k;
- Eq. (6) is enough to define a new parameter to be employed in the expander characteristic curve representation (3).

So, Eq. (3) can be expressed as (15).

$$\varepsilon, \eta, \frac{\Delta T_0}{T_{0in}} = F\left(\frac{\dot{m}\sqrt{RT_{0in}}}{P_{0in}} \middle/ \dot{m}_{corr\_crit}, \frac{N}{\sqrt{kRT_{0in}}}, k\right) \quad (15)$$

This result is even more accurate for the case in which $\mathrm{vol\_out}$ = const, that is when a nozzle is used to expand and drive the flow towards the rotor inlet. In such a case, Z variations, within the assumptions ofFig. 10 and Table 1, are negligible.

Now, let us take the second and ultimate step: applying the k influence to the resultant parameters separately (left side of (15)).

There are three resultant parameters to be calculated:

- ε, expansion ratio. This is not directly determined by the expander but some other device upstream (e.g. compressor) = > this is a boundary condition, so an input for the actual device. The expansion ratio can be used for determining the mass flow, which can be obtained once the value of the flow parameter defined in Eq. (6) is known from the characteristic curve.

- $\Delta T_0/T_{0in}$, temperature drop. This can be univocally determined from the expansion ratio and efficiency.

- η, efficiency (usually total-to-total isentropic efficiency). It is now evident that this is the only parameter where a separate influence of k must be applied.

In order to understand the influence of k over η, let us consider its definition:

$$\eta = \frac{W}{c_p T_{0in}\left(1 - \frac{1}{\varepsilon^{\frac{k-1}{k}}}\right)} \quad (16)$$

where W represents the real specific work obtained from the expander.

After the passages extensively showed in [3], η can be expressed as in (17).

$$\eta = \text{const}\frac{\sin\,\alpha_{vol\_out}}{\left(1 - \frac{1}{\varepsilon^{\frac{k-1}{k}}}\right)}\left(Ma_{rot\_in}\frac{1}{\sqrt{f}}\right)\left(\frac{N}{\sqrt{kRT_{0in}}}\right)(k-1)$$

(17)

When two points are similar, the flow coefficient in (15) has the same value, so Mach is also the same. The same reasoning is also valid for the speed coefficient. So, the expansion ratio is also the same.

Moreover, Table 1 shows how little are the variations in $\alpha_{vol\_out\_crit}$ caused by k variations: it is straightforward to calculate a variation of -0.5% for $\sin\alpha_{vol\_out\_crit}$. This implies that, for different $k, \sin\alpha_{vol\_}$ outin (17) can be regarded as a constant.

Two different values of η for two similar points (k and k›) can be obtained:

$$\frac{\eta}{\eta'} = \frac{\left(1 - \frac{1}{\varepsilon^{\frac{k'-1}{k'}}}\right)\sqrt{\frac{k'+1}{2}}\,(k-1)}{\left(1 - \frac{1}{\varepsilon^{\frac{k-1}{k}}}\right)\sqrt{\frac{k+1}{2}}\,(k'-1)}$$

(18)

where the function f has been evaluated for $Ma = 1$.

Eq. (17) represents the correction to be applied to η due to a variation in k.

Eventually, Eq. (19) is the final representation of turbine performance, where η› is the corrected value of η, which is obtained from the characteristic curves.

$$\varepsilon, \eta', \frac{\Delta T_0}{T_{0in}} = F\left(\frac{\dot{m}\sqrt{RT_{0in}}}{p_{0in}}\bigg/ \dot{m}_{corr\_crit}, \frac{N}{\sqrt{kRT_{0in}}}\right)$$

(19)

For the sake of completeness, in the case analysed in Fig. 10, the variation in k of -15% corresponds to a variation in efficiency of about -17% (with $\square = 4$). This demonstrates the validity for the assumption of negligible variations of $\sin\alpha_{vol\_}$ out.

Sensitivity of Fluid Properties to Chemical Composition Variations

In order to understand which are the actual composition variations that should be expected in the field, two situations are analysed (Table 2):

- I_{air}: air with 10% mass fraction of steam (corresponding to a real case of steam injection)
- I_{FCex}: typical exhaust gas composition coming from a fuel cell

Table 2: Flow compositions (mass fractions) of the two composition cases, including the air composition used for reference

Substance	Reference Air	
	Mass frac.	Molar frac.
N2	0.7504	0.7729
O2	0.2299	0.2073
H2O	0.0063	0.0101
CO2	0.0004	0.0003
Ar	0.0127	0.0092

Substance	I_{air} – Air with 10% steam mass fraction		I_{FCex} typical FC exhausts	
	Mass frac.	Molar frac.	Mass frac.	Molar frac.
N2	0.6797	0.6625	0.7300	0.7350
O2	0.2083	0.1777	0.1100	0.0970
H2O	0.1000	0.1515	0.0700	0.1096
CO2	0.0004	0.0002	0.0800	0.0513
Ar	0.0115	0.0078	0.0100	0.0070

As a result, it is possible to state that the parameters of influence, which clearly account for composition variations in the working fluid of an expander, are, in order of importance:

- c_p, specific heat at constant pressure
- R, gas constant
- k, specific heat ratio

This is the reason why, in common practice, k variations are often neglected. Nevertheless, to improve the accuracy of the calculations it is important to take these into account, especially in hybrid systems (Table 3).

Table 3: Properties of the different flow compositions (specific heats are calculated at 1000 K). Changes are evaluated against the corresponding value of the reference air

Property	Reference Air	
	Value	Change
cp [J/kgK]	1148.2	0%
R [J/kgK]	288.14	0%
k	1.335	0%

Property	I_{air} – Air with 10% steam mass fraction		I_{FCex} typical FC exhausts	
	Value	Change	Value	Change
c_p [J/kgK]	1255.9	+9.4%	1235.9	+7.6%
R [J/kgK]	304.45	+5.7%	294.73	+2.3%
k	1.320	−1.1%	1.313	−1.7%

TURBOMACHINERY IN SOFC HYBRID SYSTEMS: DESIGN POINT STUDY

SOFC hybrid systems are not commercially available yet, even if they are subject of intensive research and development efforts worldwide since several years [7]. Hybrid system performance has to be investigated with models, which have been only partially verified against experiments. Thus, experimental validation of models was not performed on the layouts studied but on portions of layouts and/or single components: results of past validation activities can be found in [3], [8] and [9]. Overall, it can be stated that on-design and off-design models are capable of predicting real plant data within an accuracy of 3%: considering an average uncertainty of field measurements in the order of 1%, this can be considered an acceptable level of confidence.

Fig. 12 shows an hybrid plant where the fuel cell works under atmospheric conditions and the energy content of the exhaust gases is recovered in a gas turbine downstream the fuel cell stack. The gas turbine and the fuel cell are linked by a high temperature heat exchanger, similar to that used in externally fired gas turbines, to recuperate the thermal energy of the fuel cell exhausts and heating up the compressed air before entering the expander.

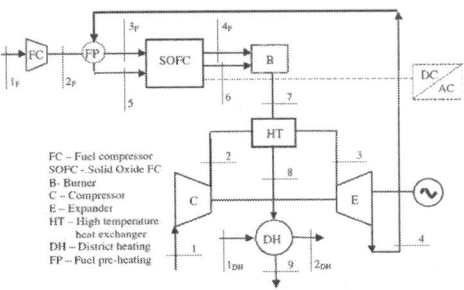

Figure 12: Atmospheric fuel cell gas turbine Hybrid System (ATM-HS).

This component works under unfavourable conditions (different pressures between the hot and the cold streams and a high operating temperature) and must be very effective in order to achieve the suitable turbine inlet temperature and keep the gas turbine at efficient operating points. These requirements make this heat exchanger critical and very expensive.

Fig. 13 shows a pressurised hybrid plant where the fuel cell substitutes the combustion chamber of a recuperated micro gas turbine. In this way the fuel cell is pressurised by the gas turbine compressor and the exhausts are directly sent to the expander avoiding the employment of the high temperature heat exchanger. Moreover the fuel cell shows a strong increase in its performance working under pressurised condition [10], [11], [12] and [13], as well as lower entropy generation [14].

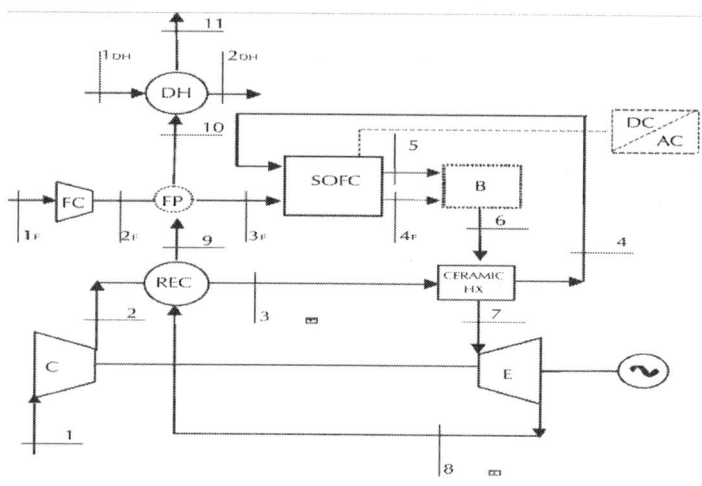

Figure 13: Pressurised fuel cell gas turbine Hybrid System (P-HS).

The performance of the pressurised Hybrid System has been investigated, initially, at design point conditions. Different sizes and types of gas turbines, e.g. radial or axial expander, have been considered in order to find the best GT configuration for the coupling with the solid oxide fuel cell stack. Following the GT definition, the SOFC group has been chosen as follows: it is fed with the same air flow rate, operates at the GT pressure, and must have the exhaust temperature similar to the TIT. To obtain such a result the fuel cell stack has been scaled-up changing the cell surface and maintaining the same current density, voltage and fuel cell temperature as design point conditions. In the case study, several gas turbines have been considered with the same TIT but with different mass flow rate, pressure ratio and expander configuration (radial or axial) in the 12 MW system power output. In terms of plant efficiency, keeping constant the current density, fuel utilization factor, fuel cell operating temperature and TIT, the most important parameter is the pressure ratio, as shown in Fig. 14.

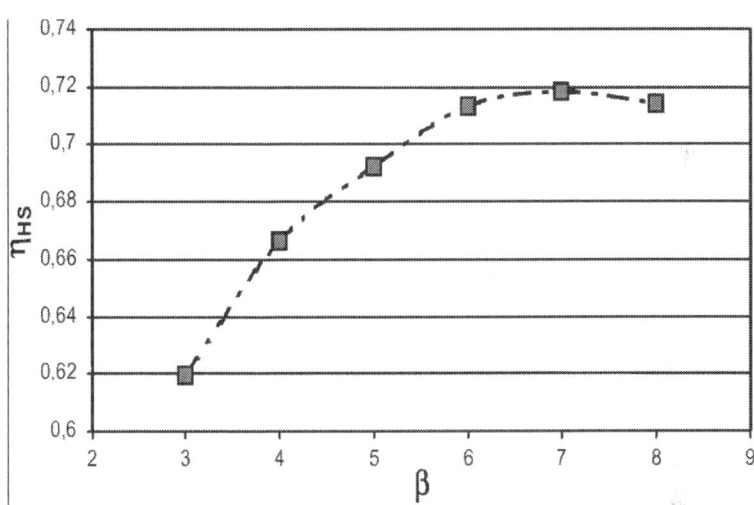

Figure 14: Hybrid system efficiency versus pressure ratio.

A maximum plant efficiency of 72% has been obtained for different gas turbine sizes at pressure ratio 7. In the case of the gas turbine sizes studied, compressor and turbine efficiencies are similar so the plant efficiency is not strongly affected by the mass flow rate. However, the choice between the radial or axial solution is influenced by considerations on cost and life for both turbomachinery and system. The power ratio between GT and fuel cell is about 25%. This depends on the GT pressure ratio, as shown in Fig. 15. It is worth noting that: (i) the gas turbine contribution is greater, taking into account that the amount of power necessary for the pressurisation of the Fuel Cell is totally charged to the GT expander, (ii) the system requires a low effectiveness ceramic heat exchanger limiting the influence on system performance and cost; (iii) the pressurised configuration avoids heavy constraints on heat exchangers in terms of temperatures, size and cost, in contrast with what has been discussed for atmospheric solutions.

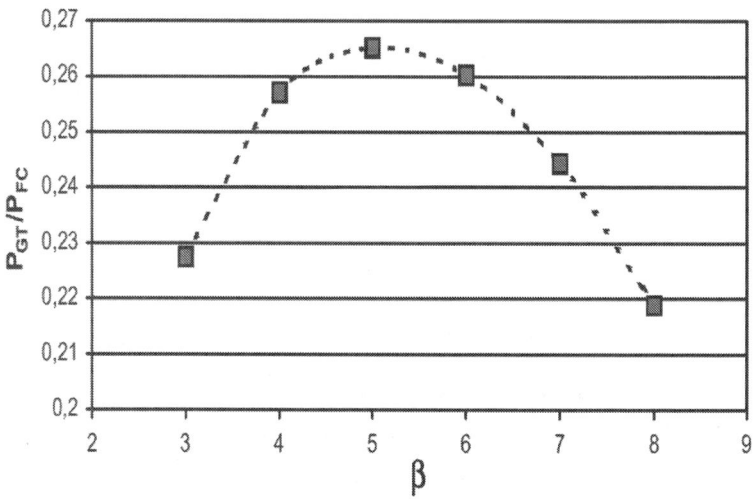

Figure 15: Gas turbine to fuel cell power ratio versus pressure ratio.

Comparison between Atmospheric and Pressurised Hybrid Systems

Fig. 16 shows the specific power [kWe/(kg/s)$_{air}$] versus pressure ratio for different plants. In particular, pressurised Hybrid System (P-HS), atmospheric HS (ATM-HS), atmospheric fuel cell (ATM-SOFC), simple cycle gas turbine (GT), and regenerated micro gas turbine (MGT).

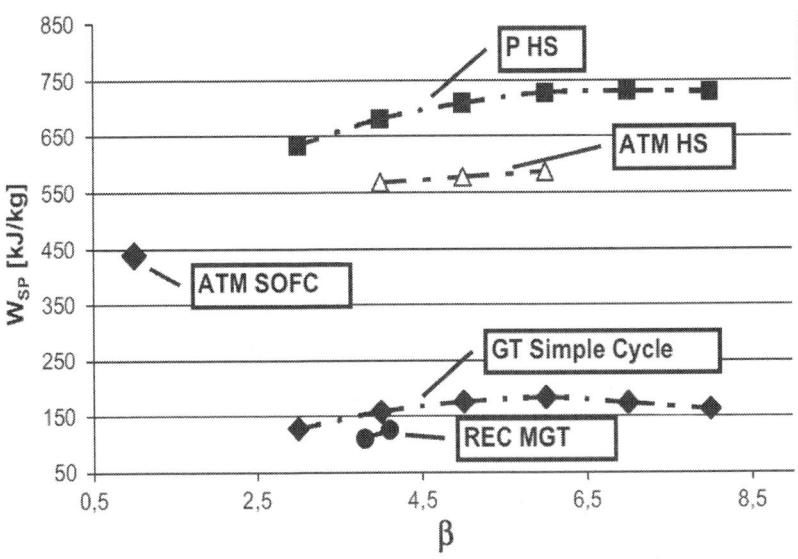

Figure 16: Specific work versus pressure ratio.

It is worthwhile noting the very high value of the specific power of the P-HS compared to the other plants. This is mainly due to the very high efficiency of the pressurised fuel cells. In addition, the high value of specific work may determine a more compact hybrid system design. Moreover, pressurised systems present less critical components than atmospheric plants. The main constraints are related to the compatibility between the characteristics of the exhaust gases of the Fuel Cell stack and the flow properties required for the turbine. In particular, the water content in the expanding flow is higher than the typical value for standard turbomachinery. However, this aspect may be easily addressed as already done for steam injection or humid cycle. In addition, the turbine inlet temperature must agree with the technological limit for small size uncooled gas turbines (about 900–950 °C).

Finally, it is worth noting that the efficiency of the pressurised system can be more than 10% higher than that of the atmospheric one, with significant advantages also in terms of specific work (Fig. 14). This aspect is very important when considering the production energy cost [15].

The main result of this comparison is the demonstration of the evident superiority of the pressurised configuration over the atmospheric one, in terms of efficiency, specific power, and cost of electricity. Moreover, the technological constraints are less stringent in the pressurised case, whilst in the atmospheric case it is very difficult to comply with the constraints on the high temperature heat exchanger. A high temperature heat exchanger is also present in the pressurised system (Fig. 13), but in this case a heat exchanger with low effectiveness is acceptable. In fact, the aim of this component in the pressurised system is mainly to pre-heat the air before it enters the fuel cell.

From an economic point of view the pressurised solution seems to be more effective, taking into account that, at a fixed fuel cell size, the generated power increases and therefore the specific power cost falls. For these reasons, the following considerations refer to pressurised fuel cell plants. Several sizes and types of gas turbines, e.g. radial or axial expander, can be considered to find the best Gas Turbine (GT) configuration for coupling with an SOFC stack: the decision variable is, of course, the size of the turbomachinery

Overall, it is important to note that the coupling of the SOFC and the gas turbine is not immediate: all the parameters of these devices must match very well, especially in terms of TIT, maximum allowable cell temperature, operating pressure and working fluid composition.

TURBOMACHINERY IN SOFC HYBRID SYSTEMS: OFF-DESIGN STUDY

The response of the plant and an understanding of the phenomena that drive it are very important, even in a qualitative way, especially for the choice of the control system. For these reasons the following off-design study of a hybrid plant based on a tubular SOFC was carried out using a simplified fuel cell model [10].

Starting from the design point defined below, the power of the plant was modified by applying two different control systems:

- control of the fuel flow rate (and simultaneous control of the electrical current of the fuel cell);
- simultaneous control of the fuel flow rate and of the rotational speed of the gas turbine.

The operating condition chosen for the design point of the HS plant was so that the turbine flow rate, rotational speed, pressure, turbine inlet temperature were compatible with the limitations of this technology and with the size of the machine under consideration. The definition of the gas turbine imposed restrictions on the SOFC system, which had to be fed with the same air flow rate, operate at the Gas Turbine (GT) pressure, and have an exhaust temperature value equal to the turbine inlet temperature (TIT) of the Gas Turbine.

The system considered in this section is for a Micro Gas Turbine (MGT) with a size of about 50 kW_e and an SOFC stack with an overall active area of 95 m^2, producing about 240 kW of power at the design point (the cell specific power is about 2.5 kW/m^2).

Even if the best operating condition for the combined plant is slightly different from the design point chosen on the basis of the previous procedure, the main goal here is the study of the behaviour of a hybrid system under off-design conditions, and this would only be marginally affected by the choice of the design point.

In the case study, the typical operating conditions are as follows: pressure 0.38 MPa, fuel (methane) flow rate and inlet temperature 0.0094 kg/s, 15 °C, oxidant (air) flow rate and inlet temperature 0.47 kg/s, 15 °C, respectively.

The temperature, pressure and flow rate calculated for each point of the plant at the HS design point are shown in Fig. 17.

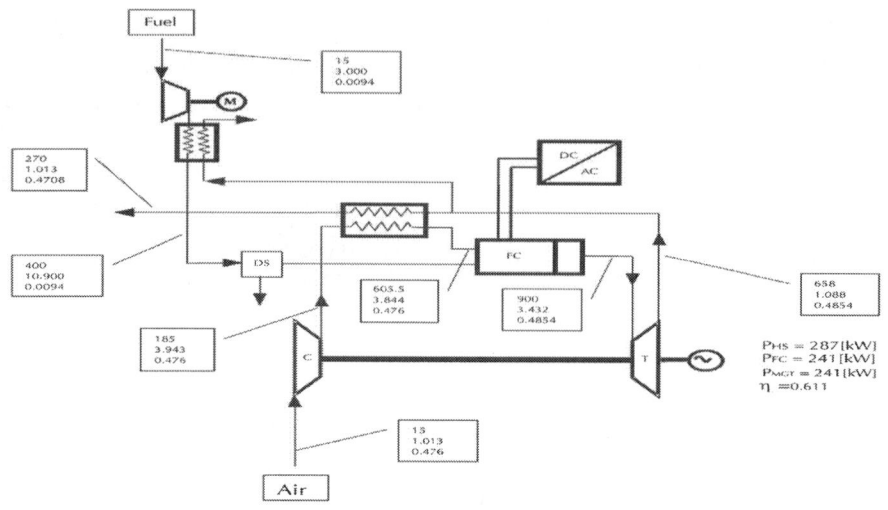

Figure 17: Pressurised SOFC hybrid system.

It is interesting to note that the HS efficiency is near to 60%, and this result is particularly important when taking into account the small size (287 kW$_e$) of the whole plant. This efficiency is higher than the efficiency of large-size advanced combined cycles (>200 MW) using steam as blade cooling media.

It is also important to note that under design conditions the MGT power is about 16% of the whole HS power, while the ratio between MGT and SOFC power is 19%. However the design point performance is not sufficient to determine whether the HS is a good system for the distributed electricity market and cogeneration applications.

In fact, it is possible to imagine that the HS will operate at part-load conditions for long periods of its operating life.

Constant Rotational Speed

If a variable speed control system is not available, the only way to vary the power supplied by an MGT plant (for part-load operation) is to vary the overall fuel flow rate. In all the calculations, the cell fuel utilisation factor has been kept constant at 0.85, and thus different fuel flow rates correspond to proportionally different electrical currents supplied by the SOFC stack. On the other hand, the air flow rate is an independent parameter, which is evaluated on the basis of the matching of the MGT expander, the MGT compressor, and the stack, thus the oxygen utilisation is not constant throughout the simulation.

Thus, the power consumption of the air and fuel compressors and the mechanical losses are not taken into account in the efficiency of the fuel cell system, while they are considered in evaluating the efficiency of the overall plant. To give an idea of the relative importance of all the contributions, at the design point the fuel cell and the MGT expander supply 242 and 136 kW respectively, the air compressor consumes 82 kW and the methane compressor and mechanical and electric efficiency account for 9 kW overall. Thus, as the compressors serve the SOFC group as well as the turbine, it would be realistic to subtract a part of their power consumption from the power supplied by the SOFC group in evaluating SOFC efficiency. In other words, the increase in efficiency due to the presence of the microturbine is not simply the difference between the overall plant and the SOFC system efficiencies (which could erroneously be calculated from as about 10 percentage points).

Fig. 18 shows the response of the overall plant and the SOFC group to a variation in the fuel flow rate (from the design point (DP) to the part-load conditions A, B,...G). It is important to point out that a direct comparison between SOFC and HS efficiencies is not possible as they are evaluated with slightly different procedures. In fact, the efficiency of the plant is calculated as the ratio between the net power produced and the LHV of the feeding fuel, while the efficiency of the stack is the ratio between the power supplied by the stack alone and the LHV of the fuel supplied to the plant.

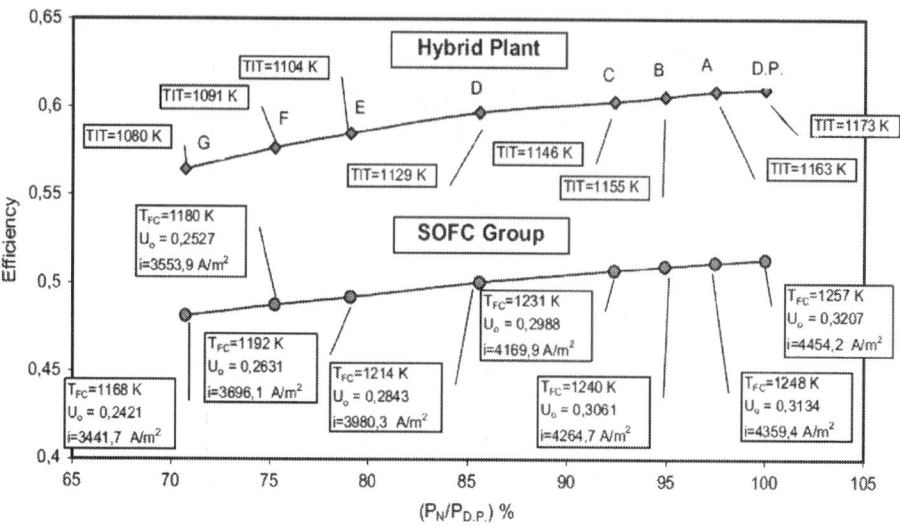

Figure 18: Off-design performance of P-HS (constant speed).

Thus, the power consumption of the air and fuel compressors and the mechanical losses are not taken into account in the efficiency of the fuel cell system, while they are considered in evaluating the efficiency of the overall plant. To give an idea of the relative importance of all the contributions, at the design point the fuel cell and the MGT expander supply 242 and 136 kW respectively, the air compressor consumes 82 kW and the methane compressor and mechanical and electric efficiency account for 9 kW overall. Thus, as the compressors serve the SOFC group as well as the turbine, it would be realistic to subtract a part of their power consumption from the power supplied by the SOFC group in evaluating SOFC efficiency. In other words, the increase in efficiency due to the presence of the microturbine is not simply the difference between the overall plant and the SOFC system efficiencies (which could erroneously be calculated fromFig. 18 as about 10 percentage points). However, it is not a simple matter to determine how to share the power losses between the turbine and the SOFC system.

Fig. 18 shows that the oxygen utilisation factor varies heavily, decreasing from 0.32 to 0.24 when the power percentage changes from 100% to 70%. The reason for the air utilisation factor varying is related to the operating conditions of the MGT compressor. In fact, at constant rotational speed, the variation range of the compressor flow rate is rather narrow and thus the air flow rate is almost constant, resulting in a sensitive variation to the oxygen utilisation at different current densities. This oxygen utilisation variation is the cause of the stack temperature varying from 984 °C to 895 °C, and thus the efficiency varying from 52% to 48%.

The efficiency vs. power curve of the plant is parallel to the efficiency vs. power curve of the SOFC system (Fig. 18), which means that the power supplied by the MGT, minus all the power losses, is directly proportional to the fuel flow rate through all these simulations. The off-design variation of efficiency of the HS is less than 5% under the conditions discussed so far, varying from 61% at the design point to 56.4% at 70% of the nominal power (the minimum part-load conditions considered here are determined by the constraints on the MGT expander inlet temperature, about 820 °C).

In order to explain the efficiency vs. power curve of the SOFC system in greater detail, Fig. 19 shows the voltage and electrical resistance as a function of the electrical current density for each of the operating points previously reported in Fig. 18, where the operating temperatures are reported as well. Again, the decrease in the cell voltage and its departure from the Nernst potential when the electrical current density is decreased is an unexpected result. The increase in the electrical resistance by decreasing the electrical current density is the explanation, and it is due to the notable decrease in temperature, which is reported in the labels to Fig. 18.

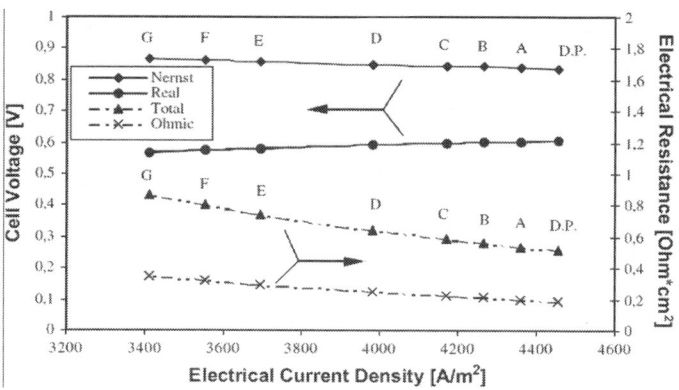

Figure 19: Off-design performance of fuel cell stack in P-HS layout (constant speed).

Variable Rotational Speed

The typical operation mode of large-size gas turbine plants does not usually involve the possibility of changing the rotational speed of the turbine. The reason for this is that typical plants do not include an inverter, and thus the rotational speed of the turbine is chosen on the basis of the alternate current frequency required by the end user/electrical network. On the contrary, an HS requires the presence of an inverter which converts the electrical current produced by both the fuel cell and the alternator, the latter being rectified previously, to direct current. Thus, this configuration allows the operation of the turbine at variable rotational speed (variable frequency).

The part-load performance of the hybrid plant at different turbine rotational speeds is depicted in Fig. 20. The results show many interesting features:

- the overall efficiency of the system is very high, as expected;
- all the fixed turbine rotational speed curves show an average decrease in plant efficiency of about 0.05 if the load is reduced by 30%. Only the curves at low turbine rotational speed show a slightly higher loss of efficiency at part-load conditions;

- the possibility of varying the rotational speed of the turbine is of fundamental importance for operating the plant at very high efficiency, even at very low part-load conditions. For example, Fig. 20 shows that at a load of about 35% of the nominal operating power, the plant efficiency is about 52% at 65,000 rpm; in addition, it is interesting to see that at 75% of the nominal operating power the efficiency increases from 57.6% to 60% to 61.7% by decreasing the rotational speed from 85,000 to 80,000 to 75,000 rpm respectively. With reference to Fig. 20, it is interesting to point out here that, for each turbine rotational speed, the overall plant has been simulated between a maximum load obtained when the turbine inlet temperature is maximum (900 °C), and a minimum load reached when the turbine inlet temperature is about the minimum operating value (about 820 °C);

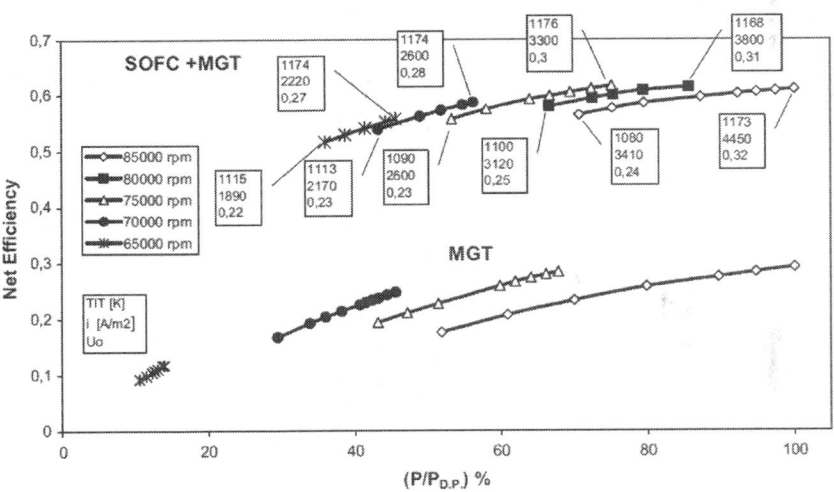

Figure 20: Off-design performance of P-HS (variable speed).

The advantage of the hybrid system over a traditional MGT plant at both the design point and under part-load conditions is evident from Fig. 20.

In order to investigate the performance of the different parts of the plant at part-load conditions in further detail, attention was limited to the SOFC system itself. The results are reported in Fig. 21 and show that for each fixed value of the turbine rotational speed the behaviour of the SOFC system is very similar to that previously reported in Fig. 20 for the overall plant.

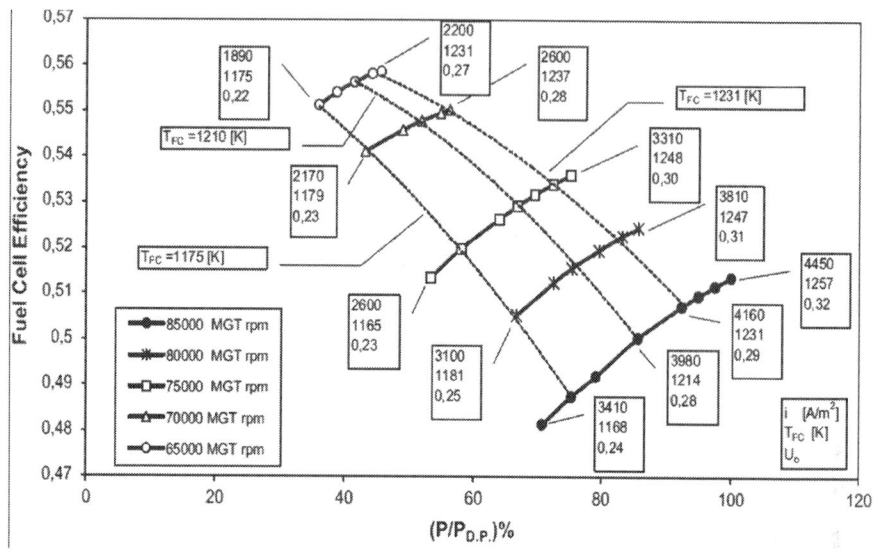

Figure 21: Off-design performance of fuel cell stack in P-HS layout (variable speed).

The interesting feature reported in this figure is that the fuel cell efficiency increases when the turbine rotational speed is decreased, while the opposite is true for the overall plant. It is possible to give an explanation for this effect by considering the broken curves in Fig. 21, along which the efficiency increases when the power is decreased. It can be noticed (labels to Fig. 21) that along each broken curve both the temperature and the oxygen utilisation of the fuel cell remain fairly constant; this is due to the fact that, when the SOFC module is integrated into the plant, there is an interesting effect on the recuperator performance when the variable speed control system is utilized. In fact (labels to Fig. 21), in this case the load change involves a reduction in the air flow correlated to the speed reduction, while for fixed speed control the load reduction is only due to the fuel mass flow rate variation. Therefore, under variable speed control conditions, the recuperator effectiveness increases significantly at part-load with decreasing air flow, until a very low value is reached (below 10%), where thermal performance is degraded by longitudinal conduction in the heat exchanger, particularly in the case of highly compact matrices. This variation in recuperator effectiveness allows the air temperature at the inlet of the stack (recuperator outlet) to increase at part-load, and therefore the SOFC temperature remains fairly constant along the broken lines of Fig. 21.

Since both the temperature and oxygen utilisation factor remain constant along those curves, they show an increase in fuel cell efficiency when the power is decreased. In spite of the increased performance of the SOFC stack under part-load conditions, the overall plant shows a decrease in performance under the same conditions (Fig. 20), which is due to the fact that the efficiency of the fuel cell system alone does not take the power required by the compressor and the off-design of the electrical part of the plant into account, as already mentioned. These effects, and also all the effects relating to the off-design of the MGT group, the alternator (efficiency of 85% at 25% load) and the mechanical parts, explain the difference between Fig. 20and Fig. 21.

CONSIDERATIONS ON CONTROL SYSTEM

As a final note on the dynamic behaviour of hybrid systems, both MCFC and SOFC based, it is useful briefly discussing the main specifications and constrains the control system has to comply with[16] and [17].

A hybrid system cannot be considered, in its present stage of development, as quick a power plant as a gas turbine, even if turbomachinery is present. In fact, unless the gas turbine is capable of running on its own, so implying the presence of a very hot valve somewhere in the fuel cell exhaust duct, the hybrid can hardly be employed for peak electricity shaving. It seems, instead, more suitable to penetrate the market for highly-efficient distributed generation of a base-load, wherever electricity is needed at quite a constant level for long periods of time (alternatively, electricity can always be sold to the public grid, with, however, a decrease in the revenue from the plant). In this case, the physical characteristics of the cycle and the control system embedded in it become the main factors that can determine the success of fuel cell hybrid systems.

The first task for the control system is the management of the start-up and shut-down phases of the hybrid. Here the main specification is, of course, the time lag between the start and the end of the procedure, which needs to be as fast as possible for increasing the plant flexibility. Unfortunately, the physical constraints of the hybrid system are likely to strongly affect the total time required. The main parameters to be monitored are:

- thermal gradient (local and average) of the fuel cell stack;
- the pressure difference between anode and cathode sides;
- the composition of the natural gas entering the fuel cell ducts;
- shaft overspeed;
- compressor surge.

The first constraint, the one that most certainly delays the system response, can be controlled by using an apt start-up combustor during start-up, or compressed washing air during shut-down. In this case, a hybrid system is also likely to need a significant external source of energy (typically electricity) for safe shut-down, which is quite an uncommon situation for power plants and reduces system reliability.

The second constraint is a mechanical one due to the mechanical stress on the fuel cell pipes (if tubular FC) or plates (if planar FC). Depending on the cycle configuration, such a pressure difference is usually lowered by a fluidodynamic link between the two sides, which tends to even out the difference. Whenever a blower is present (typical for MCFC systems), the control system can directly act on that to keep the pressure of the two sides as close as possible; otherwise, such a difference can only be indirectly controlled and smooth variations at the fuel cell operating point are required.

The third constraint is mainly due to the chemically active parts of the fuel cell: the use of expensive catalysers and specific electrolytes in order to ensure a good durability of the stack and its performance provides several constraints for the safe operation of the fuel cell. In this case, depending on the initial condition of the cell, inert gases (e.g.: nitrogen) or others can be injected in order to protect the active surface, typically from the anode side.

The forth and fifth constraints are typical of turbomachinery. In this case the challenge is to control them and the fuel cell at the same time. Actually, even if the turbomachinery inertias are quite important at the very beginning of the start-up, during normal operation the turbomachinery is likely to work at almost steady-state conditions, due to the relatively smooth response of the fuel cell system.

The second main task of the control system is the regulation of the load. Since most of the power comes from the fuel cell, it is clear that an effective load control can be pursued only by modifying the operational point of the fuel cell. Even if the electrochemical reactions and the electrical response of the cell are quite fast, so suggesting a quite prompt load-following, the constraints provided by the overall fuel cell system (especially in terms of gas composition within the fuel cell ducts) are likely to greatly limit this feature. For this reason, the first commercial generation of fuel cell hybrids will probably be equipped with some type of load dumper (e.g.: batteries, load banks, capacitors, etc...).

CONCLUSIONS

Fuel cell gas turbine hybrid systems are likely to play an important role in the clean power generation of the future. As far as high efficiency and economic viability are addressed, integration with turbomachinery is the most-studied option, because it is suitable to cope with the high temperature fuel cell requirements, mainly in terms of pressurisation and thermal management.

This paper reports about the state of the art of radial turbomachinery, which is likely to be the technology of choice for the first generation of hybrid systems to be developed by the industry. Turbomachinery modelling is analysed in details, reporting about a new comprehensive similitude theory for radial expanders, which is necessary to accurately take into account the impact of exhausts composition variation on the expander performance, thus allowing more reliable model predictions: c_p, R and k are shown to be subject to variations respect to dry air up to 7.6%, 2.6% and 1.7%, respectively. The composition influence on expander performance is investigated in details through the new analytical approach, and results are presented to demonstrate the quantitative influence of the system parameters on the performance.

The paper covers the three main aspects of performance evaluation: the on-design, the off-design and, as a final mention, the control of the fuel cell hybrid systems. Main conclusions are:

- pressurised hybrid systems are superior respect to atmospheric ones because of attainable efficiency, system compactness and economics, technological issues (possible avoidance of high temperature heat exchangers);

- the ideal pressure level for fuel cell operation within a high temperature fuel cell gas turbine hybrid system is in the 4–8 bar range;

- pressurised hybrid systems may show quite a stable efficiency at part-loads, superior to any other conventional power generation technology, when options to reduce the air flow through the compressor are possible (e.g.: variable speed microturbine);

- first generation of hybrid systems are likely to not be able to quickly respond to load variations, such as gas turbines, because of the several thermal, pressure and chemical constraints provided by the fuel cell stacks; they will rather characterise as base-load devices with mid-term load-following capability.

ACKNOWLEDGEMENTS

The support of the European Commission along the recent years to the research activity on fuel cell hybrid systems through contracts IM-SOFC-GT (ENK5-CT-2000-00302), FELICITAS (TIP4-CT-2005-516270), METHAPU (FP6-031414), BIOSOFC (MTKD-CT-2006-042436), EnSOFC (MTKI-CT-2006-042298), LARGE-SOFC (IP-019739), is greatly acknowledged.

REFERENCES

1. McDonald CF. Recuperator considerations for future higher efficiency microturbines. Applied Thermal Engineering 2003;23:1463–87.

2. Traverso A, Massardo AF, Roberts RA, Brower J, Samuelsen S. Gas turbine assessment for air management of pressurised SOFC/GT hybrid systems. Journal of Fuel Cell Science and Technology 2007;4:373–83.

3. Traverso A. TRANSEO: A new simulation tool for transient analysis of innovative energy systems, Ph.D. Thesis, TPG-DiMSET, Universita` di Genova, Genova, Italy, 2004.

4. Dixon SL. In: Fluid mechanics and thermodynamics of turbomachinery. 4th ed. Butterworth-Heinemann (BH); 1998.

5. Cohen H, Rogers GFC, Saravanamutto HIH. Gas turbine theory. Addison Wesley Longman Ltd; 1996.

6. Vavra MH. Aero-thermodynamics and flow in turbomachines. New York: R.E. Krieger Pub. Co.; 1975.

7. Layne A, Samuelsen S, Williams M, Holcombe N, Hybrid fuel cell heat engines: recent efforts. In: Proceedings of ASME Turbo Expo` 2001, 2001-GT-0588.

8. Traverso A, Massardo AF, Scarpellini R. Externally fired micro-gas turbine: modelling and experimental performance. Applied Thermal Engineering 2006;26:1935–41.

9.   Ferrari M, Pascenti M, Massardo AF. Ejector model for high temperature fuel cell hybrid systems: experimental validation at steady-state and dynamic conditions. Journal of Fuel Cell Science and Technology 2008;5(041005):1–7.

10.   Magistri L., Hybrid Systems for Distributed Generation, Ph.D. Thesis, TPGDiMSET, Universita` di Genova, Genova, Italy, 2003.

11.   Liese EA, Gemmen RS. Performance comparison of internal reforming against external reforming in a solid oxide fuel cell, gas turbine hybrid system. Journal of Engineering for Gas Turbines and Power 2005;127:86–90.

12.   Harvey SP, Ritcher HJ. Gas turbine cycles with solid oxide fuel cells. Part 2: A detailed study of a gas turbine cycle with an integrated internal reforming solid oxide fuel cell. Journal of Energy Resources Technology 1994;116(4):312–8.

13.   Leo AJ, Ghezel-Ayagh H, Sanderson R, Ultra High Efficiency Hybrid Direct Fuel Cell/Turbine Power Plant. In: Proceedings of ASME Turbo Expo` 2000, pp. 2000-GT-0552.

14.   Sciacovelli A, Verda V. Entropy generation analysis in a monolithic-type solid oxide fuel cell (SOFC). Energy 2009;34(7):850–65.

15.   Massardo AF, Magistri L. Internal reforming solid oxide fuel cell gas turbine combined cycles (IRSOFC-GT) – Part II: exergy and thermoeconomic analyses. Journal of Engineering for Gas Turbines and Power 2003;125(1):67–74.

16.   Ferrari M, Magistri L, Traverso A, Massardo AF, Control System for Solid Oxide Fuel Cell Hybrid Systems. In: Proceedings of ASME Turbo Expo` 2005, pp. GT2005–68102.

17.   Kandepu R, Foss LI, Stiller C, Thorud B, Bolland O. Modeling and control of a SOFC-GT-based autonomous power system. Energy 2007;32(4):406–17.

Chapter 8

Fluid Mechanics Research at the Institute of Aerodynamics, RWTH Aachen University: From 1912 through 2012

W. Schröder

Institute of Aerodynamics, RWTH Aachen University, Wüllnerstraße 5a, 52062 Aachen, Germany

INTRODUCTION

In the year 2012 the Institute of Aerodynamics celebrates its 100th anniversary, i.e., 100 years of research in the field of fluid mechanics. Without any doubt it is impossible to give a detailed survey of the research done at the Institute of Aerodynamics over the last 100 years in just a few-page article. For this reason, only an excerpt of the various studies is presented, i.e., the following discussion is by no means complete, it is supposed to give just a flavor of the scientific breadth that has been covered. The reader who is interested in more details will find an exhaustive publication list in [1], [2], [3] and [4]. This concise presentation of the history of the Institute of Aerodynamics is based on the following structure. Each period of the various directors of the Institute of Aerodynamics is considered separately. First, the scientific vita of the director is briefly discussed and then, some scientific highlights achieved during his directorship are succinctly outlined. This decomposes the 100-year history into six subsections.

THE PERIOD FROM 1912 TO 1931

At the beginning of the 20th century Fluid Mechanics became an independent scientific discipline. Especially Aerodynamics boosted the research in Fluid Mechanics. At Aachen in these days Aerodynamics research was strongly supported by Prof. H. Junkers and Prof. H. J. Reissner (Fig. 1).

Figure 1: Hans-Jacob Reissner.

Until 1899 Prof. A. Ritter taught Mechanics at the RWTH Aachen University. His successor was Prof. A. Sommerfeld who held the Chair of Mechanics from 1900 through 1906 before in 1906 Prof. H. J. Reissner took over the Chair of Mechanics. In 1909 Reissner and Junkers performed the first aerodynamic experiments which mark the beginning of several significant experimental and theoretical contributions to the development of aircraft. In 1912, the first all-metal aircraft, the so-called Reissner Duck was ready to fly. Several successful flights were conducted near Aachen at the Branderheide. Although Prof. H. J. Reissner accepted a call of the Technical University Berlin in 1913, there is no doubt that it was his pioneering work in Aerodynamics and his scientific stimulus that made the idea of an Institute of Aerodynamics and Chair of Fluid Mechanics come true at RWTH Aachen University. Prof. H. J. Reissner was still at Aachen when the foundation stone ceremony of the Institute of Aerodynamics took place in 1912.

On April 1st, 1913 Prof. Th. von Kármán (Fig. 2) took over the Chair of Mechanics and Aerodynamics and became the first Director of the Institute of Aerodynamics. The name of the Institute underlined the scientific significance of this new research discipline. The building up of the scientific reputation of the Institute of Aerodynamics is definitely his accomplishment. He collaborated closely with Prof. E. Trefftz who held a Professorship in Mathematics at Aachen and with Prof. L. Hopf who was a scientist at the Institute of Aerodynamics before he became a Professor of Mechanics at Aachen in 1923. From 1926 on v. Kármán spent part time at Aachen and Pasadena in California where in 1930 he became the Director of the Guggenheim Laboratory at the California Institute of Technology. On 1 April 1934 v. Kármán left Aachen for good and moved to Pasadena. However, all his life long he kept a very close relationship with the Institute of Aerodynamics. In 1957 v. Kármán became the first Prandtl-Ring recipient. Prof. H. F. G. Bock was appointed Professor of Mechanics and Aerodynamics in 1934. Unfortunately, he never really did take office since he also held a position at the Aerospace Department of the German Government in Berlin. In 1936 he accepted an executive position at the Deutsche Versuchsanstalt für Luftfahrt (DVL) in Berlin.

Figure 2: Theodore von Kármán.

The v. Kármán era at Aachen is characterized by a strong link between fundamental research and application (Fig. 3). In the following, some highlights of the research that had been performed under his guidance at the Institute of Aerodynamics are given. Note that due to lack of space this list is far from complete.

Figure 3: The Institute of Aerodynamics 1922.

Based on Prandtl's boundary-layer theory detailed research on boundary layer flow was conducted at Aachen. In a fundamental article v. Kármán rewrote the momentum equation in the streamwise direction, i.e., he formulated the von Kármán integral condition which substitutes the partial differential equation by a first-order ordinary differential equation for the boundary-layer thickness [5] and [6]. Using this integral condition Pohlhausen developed an iterative integration method [7] the concept of which is still used today in many design methods to investigate viscous flows over slender bodies. The semi-empirical formulation by von Kármán and Prandtl for turbulent friction was used in conjunction with the analogy of Reynolds and Prandtl to determine heat transfer phenomena. Assuming that momentum and heat transfer are defined by the same mechanism of arbitrary molar fluctuation v. Kármán derived the differential equation of turbulent heat conduction which made it possible to compute the heat transfer provided the turbulent velocity distribution was known. The experimental results of the temperature and velocity field at a flat plate of Eliás [8] confirmed the theoretical Prandtl–Kármán solutions. The investigation of the nature of turbulence, i.e., the exchange mechanism within the flow field, was a major research area in the 20s of the last century. Following a similarity analysis on the fluctuation mechanism v. Kármán derived a relationship between the distribution of the mean velocity and Prandtl's mixing length. This equation resulted in the formulation of a universal velocity distribution for pipe flow [9]. This universal law contains the Reynolds number in a form depending on the wall-shear stress such that it is valid for smooth and rough walls. A similar formulation described the effect of friction for a flat plate. A summary of the development of those findings is given in [10]. The stability of laminar flow and the generation of turbulence were mathematically analyzed by Hopf [11] and [12] and the scientific findings were summarized and extended by v. Kármán in [13].

The airfoil theory was further developed by v. Kármán and Trefftz [14] by computing lift and moment of airfoils whose trailing edge possessed a finite angle. This resulted in an extension of the

Zhukovsky airfoils. In [14] an approximate method for prescribed airfoil shapes is also discussed. In the following, some contributions on fundamental computing methods for propfans at minimum energy loss are mentioned. Bienen and v. Kármán [15] and [16] extended the vortex theory such that they took into account the profile drag to determine the optimum thrust distribution in the sense that the induced losses are minimized. This extension led to useful computing methods for various operation conditions of propfans [17] and [18]. Even the impact of the fuselage was taken into account on the propfan design [17]. The vortex theory, which is based on the airfoil theory, also was applied to compute axial flow turbines [19]. Furthermore, the interaction between different aircraft components such as fuselage and wing were analyzed by potential flow theory methods. These methods also were used to improve the aerodynamics of sailplanes. Sailplanes allowed a very close link between theoretical and practical work which was highly appreciated by v. Kármán. Therefore, he supported the foundation of the Flugwissenschaftliche Vereinigung Aachen (FVA) an organization that still exists. The glider Schwarze Duevel (Black Devil), which was manufactured in the Institute of Aerodynamics, flew a world record in 1920 by doubling the flight distance. In 1921 another famous glider, the Blaue Maus (Blue Mouse) (Fig. 4), also set a record of 13 minutes in total flight time. Further analyses of fluid mechanics problems were related to flight dynamics, i.e., stability and control of unsteady motions were investigated, flow resistance in shaft installations, groundwater flow, fuel atomization in carburators, etc. It is also worth mentioning that it was v. Kármán who brought Göttingen-Type wind tunnels to Kobe in Japan and Pasadena in the US.

Figure 4: The Blaue Maus (Blue Mouse) and the Schwarze Duevel (Black Devil).

Although von Kármán's Aachen research is strongly related to fluid mechanics in general and aerodynamics in particular, he also worked in the field of structure mechanics. Among other studies, he published results on elastic limit states. He postulated that based on a limit principle only this much material enters the limit state that the yield criterion between a solid and a fluid state is satisfied. This minimum principle uniquely separates elastic and ductile regimes. Computational methods of the compound effect of bending and torsional load for self-supporting wings were discussed together with Friedrichs in [20]. Even questions on how to suppress flutter of the empennage and the stability of riveted joints of thin metal plates for aircraft wings were addressed.

Prof. von Kármán ideally combined engineering, physical, and mathematical knowledge which made him a brilliant scientist. Furthermore, he was an outstanding pedagogue who was able to fascinate students by the idea of interdisciplinary science.

THE PERIOD FROM 1931 TO 1942

Before Prof. C. Wieselsberger (Fig. 5) took over the Chair of Fluid Mechanics and Applied Mathematics on 1 December 1931 and later on was appointed Director of the Institute of Aerodynamics, he had been a research assistant and group leader at the Aerodynamische Versuchsanstalt (AVA) in Göttingen from 1912 through 1922 and until 1930 a scientist with the Institute of Aerodynamics of the Imperial University of Tokyo. His work focused on measurement methods, measurement instruments, and experimental facilities and it was strongly influenced by his Göttingen time, i.e., the Prandtl school.

Figure 5: Carl Wieselsberger.

The 30s were characterized by a massive increase of the available electrical power that led to the analysis of sonic flow. Wieselsberger strongly supported the development to analyze compressible flows which is why he initialized the construction of an intermittent supersonic facility. Due to the boost in aeronautics the Chair and the Institute grew in manpower and were shifted from the Department of General Sciences to the Department of Mechanical Engineering which it is still a member of today. At the beginning of the Second World War the Institute had to be moved in just a few hours from Aachen to the AVA in Göttingen. Two years later in 1941, the Institute was relocated at Aachen. The research and the teaching were strongly defined by the unstable wartime. Due to a painful disease Prof. Wieselsberger passed away on 26 April 1941. The provisional administration of the Institute of Aerodynamics was taken over by Prof. R. Sauer who had been with the Institute when it was located in Göttingen from 1939 through 1941.

At the beginning of the 30s the subsonic wind tunnel was complemented by a six-component gauge (Fig. 6) which had been developed by Wieselsberger in Japan [21]. Thrust and moment coefficients as well as efficiency measurements of propfans were performed in the low-speed facility. Especially, the fluctuation of the blades due to vortex shedding was analyzed. Furthermore, the lift distribution as a function of wingspan near maximum lift was experimentally and numerically investigated [22] and [23].

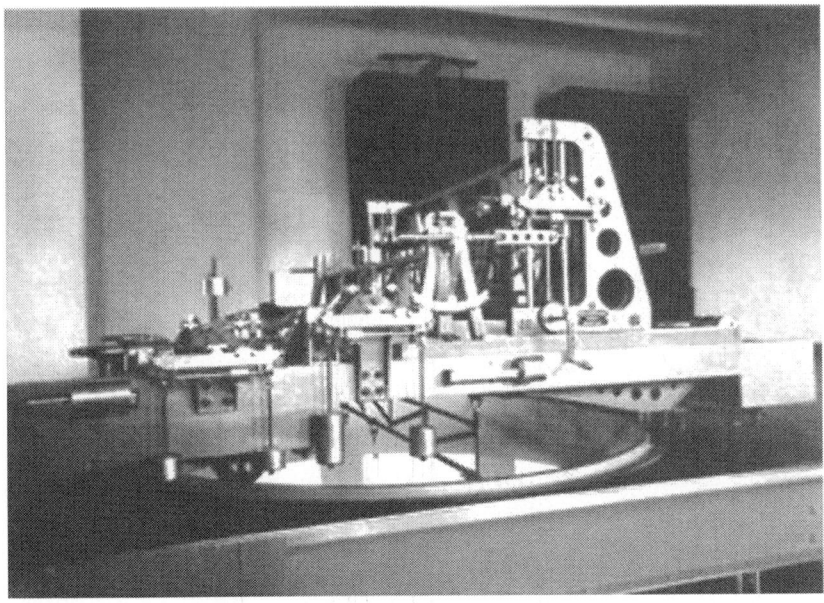

Figure 6: Six-component gauge.

Considering internal flows it was primarily the secondary flow structure in turbulent pipe flows that was thoroughly measured. In 1933 Wieselsberger designed and constructed a small supersonic wind tunnel at a test cross section of 10×10 cm^2 and a measurement time of 25 s. The nozzle shape was determined by the method of characteristics [24]. Like in other labs the X-shock phenomenon was observed near the throat region (Fig. 7). Detailed studies showed this discontinuity to be caused by the condensation of the water vapor contained in the ambient air [25] and [26].

Figure 7: Condensation shock or X-shock in the laval nozzle.

The short measurement time of such shock tunnels required a fundamental modification of the measurement technique which led to the development, test, and application of inductive measurement methods [27]. The basic principle is still used today in high-speed facilities for pressure and force measurements. There was hardly any knowledge on flows in high-speed facilities, i.e., the acceleration and deceleration of supersonic flows in a nozzle–diffusor system was not understood. Therefore, a fundamental measurement series was conducted in the Aachen tunnel. Later on a 20×20 cm^2 measurement section was installed which enabled systematic drag and moment measurements of ballistic geometries. It goes without saying that most of the military related investigations were not published but were primarily discussed in internal reports. Among other investigations, detailed measurements of the impact of the shape of the nose of bodies of revolution were conducted in the Mach number range M≈2. The minimum drag was achieved by an ogival nose shape. In further measurements the supersonic aerodynamics of the fins of the V2 rocket was analyzed. Under the guidance of Hermann the experience from the Aachen tunnel was used to build the supersonic facility in Peenemünde for the development of the V2 rocket. In another set of measurement series the heat transfer for a supersonic cone flow was investigated. In transonic flow Wieselsberger did analyze the impact of coating on free-shear layers like a free jet in a closed measurement section [28].

The results were used to minimize the impact of the wind tunnel jet on the induced drag of a wing. It is fair to state that these findings represent the basic idea of slotted walls used in transonic flow. Besides the fundamental analyses of heat transfer modes in turbulent free convection and turbulent pipe flow at high Prandtl numbers in the range of Pr=200 also the most practical shape of cooling systems of liquid cooled combustion engines of aircraft was thoroughly investigated. Furthermore, the question of the impact of wall heating on the overall drag was experimentally analyzed. The measurements showed a dramatic increase of the friction drag in laminar flow due to the enhanced instability of the boundary layer resulting in a laminar–turbulent transition at lower Reynolds numbers. In the field of structure mechanics studies on the properties of light-metal alloys at low temperatures were published [29]. The thermal expansion of various cast alloys and the changes of the microstructure were analyzed for combustion engines. A method to measure internal stresses based on the deformation of a bare hole was developed and applied to determine the internal stresses in autogenous and electric welding lines.

The research approach of Prof. Wieselsberger, who was an enthusiastic pilot, was strongly determined by the Prandtl school.

His scientific ideas and his human character were highly esteemed by his colleagues and his students.

THE PERIOD FROM 1942 TO 1963

In March 1942 Prof. F. Seewald (Fig. 8) took over the Chair of Applied Mathematics and Fluid Mechanics, which in 1948 became the Chair of Fluid Mechanics, and was also appointed Director of the Institute of Aerodynamics. From 1924 through 1936 he was a research scientist with the Deutsche Versuchsanstalt für Luftfahrt (DVL) in Berlin and later on Head of this research institution. Under his direction the DVL became an extremely important scientific research center in Germany.

Figure 8: Friedrich Seewald.

At the beginning of the 40s, the research in the Institute of Aerodynamics strongly suffered from the continuously increasing air raids. 1943 and 1944 the Institute was moved to Sonthofen in Bavaria to return not before 1947 to Aachen. Especially the ban on research in almost all areas of fluid mechanics in Germany after the unconditional surrender impeded enormously the scientific rebuilding of the Institute of Aerodynamics. This situation was aggravated by the elimination from any international research and/ or cooperation. In the late 40s and in the 50s the scientific studies focused on wave propagation and reflection, the application of this fundamental knowledge to engines, flow structures in pipes with elastic walls, and shock-wave–boundary-layer interactions with separation. The low-speed wind tunnel was available and a test bed for safety valves had been installed. In the first years after the Second World War the lectures covered general fluid mechanics in mechanical and civil engineering, heat transfer, heating and ventilation, and unsteady fluid mechanics and thermodynamics in combustion engines. Since the mid 50s the courses were allowed to address again the newest developments in aerodynamics. Prof. Seewald realized that it was necessary to conduct research in aerodynamics on a larger scale than it was possible at university institutes. It was his initiative to refound the DVL. Later he became the Honorary President of the Board of Trustees of the DVL. End of March 1963 F. Seewald became a professor emeritus.

Due to the enormous impact of the Second World War on the history of the Institute of Aerodynamics it makes sense to decompose this phase into the periods 1942 through 1945 and 1946 through 1963.

1942–1945: The work in the subsonic wind tunnel focused on pressure and force measurements for various ballistic configurations. After the Second World War this low speed facility had to be dismantled. In supersonic flow systematic measurements of the pressure distribution and the pressure gain in diverging and converging–diverging diffusors at varying the length and the angle of the diffusors were conducted. The study in [30] resulted in the relationship between the total pressure loss and the area ratio of the throat between nozzle and diffusor. Further investigations addressed the physics of shocks at separated flow and the pressure distribution at the base of blunt and conical bodies. The supersonic tunnel was confiscated by the Allied Forces after the German Surrender in 1945. The research in the heat transfer field concentrated on light-metal cooling systems. The internal flow through radiator cowlings was optimized for quite a variety of external flow conditions. The criteria and relationships to configure an optimum cooling system are summarized in [31] and [32].

1946–1963: To perform research in subsonic aerodynamic flows a new wind tunnel was installed in 1954/1955 once again on the roof of the Institute of Aerodynamics. Measurements were conducted in numerous projects in aircraft design, building aerodynamics, aerodynamic excitation of structural oscillations, and drag reduction of flat plates and/or slender bodies having specially prepared surfaces. Right after the Second World War aeronautical research was prohibited, however, the analysis of flows in piston engines was allowed. The studies of unsteady flows focused on the impact of waves from the exhaust pipes on the charge cycle and the overall performance of two-stroke engines and on the influence of wave propagation in pipe systems of injection pumps on the injection physics and the control of the injection valve. Fundamental research was performed on unsteady flows in general and the physics of flows of pulsatile engines in particular. A graphical method was developed to determine the charge cycle of two-stroke engines at large amplitude pressure waves in the pipe system connected to the engine. Fig. 9shows the paraboloid containing all the states that exist if no losses occur [33]. This paraboloid is a general representation of the state diagrams of one-dimensional flow based on the theory of characteristics.

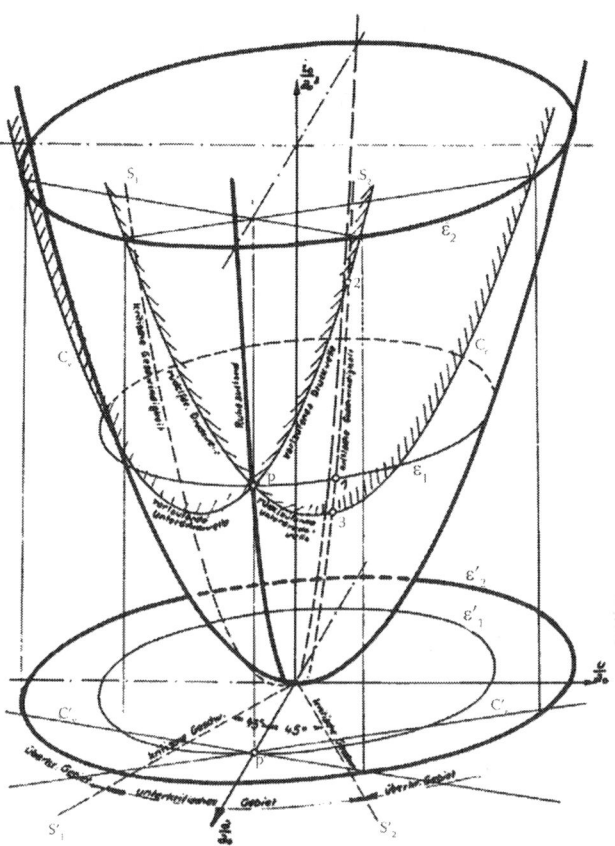

Figure 9: State paraboloid at ideal one-dimensional gas flow.

When the theoretical results of Zeller [34] on steady and unsteady flows in pipes with sudden expansion were applied to predict the charge cycle of Otto and Diesel engines, good agreement with experimental data was achieved [35] and [36]. Furthermore, the influence of the combustion chamber and exhaust pipe and the cooling system on the mode of operation of a combustion device was studied along with the possibility to control this device. Compared to steady flows, a frequency dependent enhancement of the heat transfer was observed [37]. More work on unsteady flows was done for jet apparatuses operated in pulsatile mode. In [38] the theory of unsteady one-dimensional inviscid flow of slightly compressible fluids was used to develop interference laws and reflection conditions for pressure waves in liquids. These results were the basis for a computational method to determine the propagation of waves generated by injection pumps or by the needle of injection valves. Using this theory the movement of hydraulically controlled gas valves were also computed. The propagation of pressure waves through sudden expansions and in elastic pipes and their reflection at elastic–solid boundaries were intensively studied in [39] and [40].

The measurement of the aforementioned unsteady flows required the development of new small measurement sensors. Therefore, the size of the strain gauges and the pressure sensors were reduced. Novel heat transfer data were used to optimize the size and the wire material of hot wires to increase their sensitivity and minimize the response time. Thermometers were developed to measure time dependent temperature distributions up to 4 kHz.

Research in the field of gasdynamics was forbidden until the early 50s. A new small supersonic wind tunnel at a 6×8 cm^2 measurement cross section and a maximum Mach number of 2.8 was set up in 1953 (Fig. 10).

Figure 10: Supersonic wind tunnel 1953.

A few years later, i.e., in 1963, a larger test section of 15×15 cm² extended the supersonic facility. Considering compressible pipe flow it was shown that the concept of a hydraulic diameter is valid up to a velocity range defined by the sound velocity. In the context of flow rate control at jet engines operating in pulsatile mode so-called aerodynamic throttles (Fig. 11) were investigated to determine the friction drag at standard and opposite flow direction [41].

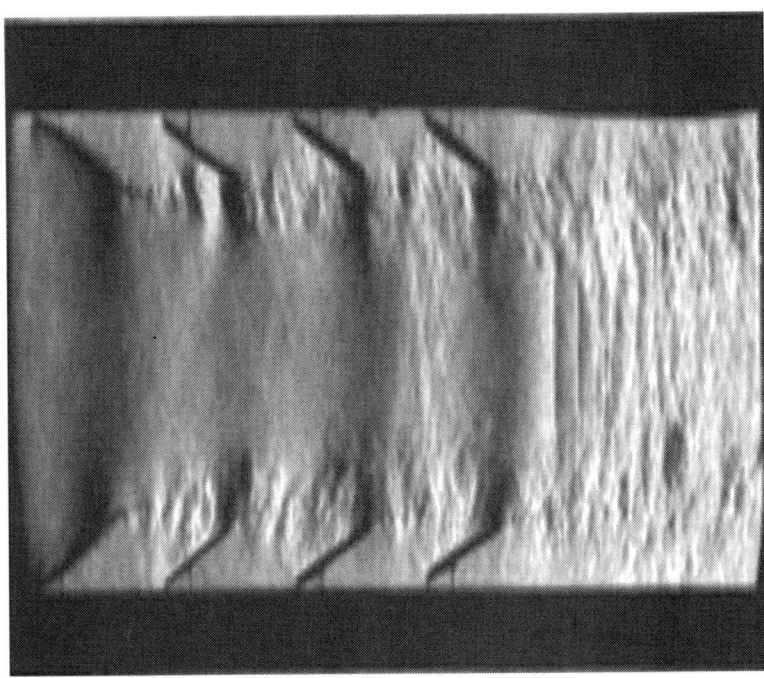

Figure 11: Schlieren picture of an aerodynamic throttle.

Moreover, detailed experiments of compressible cylinder flow were performed in the range of the critical Reynolds and Mach number to analyze the interaction of the detaching boundary layer with the generation and oscillation of local shock waves (Fig. 12). The alternating vortex shedding was shown to also occur at flow speeds higher than the critical Mach number.

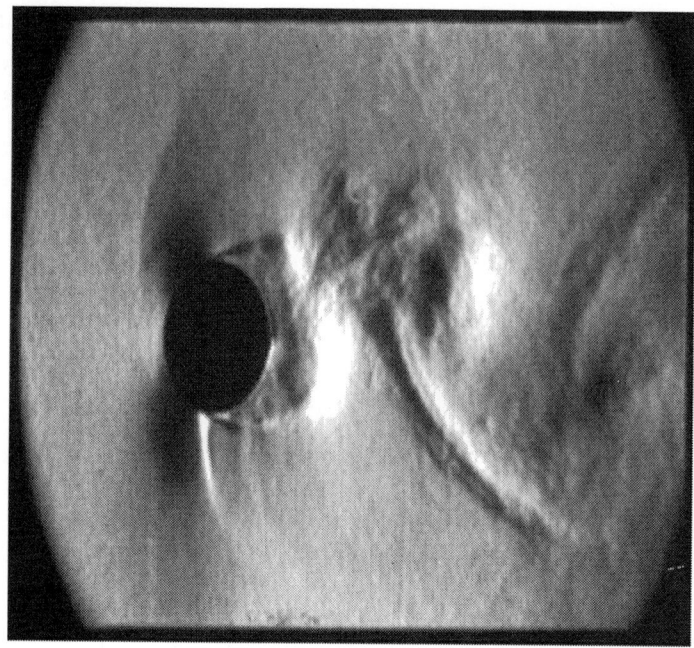

Figure 12: Schlieren picture of cylinder flow.

Besides the impact of pulsating flow on the heat transfer in heat exchangers [42] and the unsteady heat transfer in clutches [43] free convecting flow due to pronounced density gradients in ventilated rooms were investigated in [44]. The increasing significance of heat transfer research and the quality of the existing results in this research field led to the establishment of a new Chair of Heat and Mass Transfer in 1956. One year before, the Institute of Aerodynamics was asked by the Vereinigung der Technischen Überwachungsvereine (TÜV) to perform fluid mechanical tests for safety valves. This was the beginning of the development of the aerodynamic and gasdynamic fundamentals of armatures, i.e., especially safety valves.

Prof. Seewald strongly pursued the idea of reviving the DVL. In this sense, the Institute of Aerodynamics became the origin of the new DVL. In 1952 a small team started to preplan new aerodynamic test facilities. In the course of time, the idea of aeronautical research had substantially grown and the DVL had become a legal organization such that the aforementioned group separated from the Institute of Aerodynamics to become the DVL Institute of Aerodynamics and later on the DVL Institute of Applied Gasdynamics. Since the professional and personal relationship between the Institute and the DVL was very intense, many scientific ideas initialized in the Institute of Aerodynamics were further analyzed by the DVL.

THE PERIOD FROM 1963 TO 1973

In April 1963 Prof. A. Naumann (Fig. 13) became the successor to Prof. F. Seewald, i.e., he was appointed Director of the Institute of Aerodynamics and took over the Chair of Fluid Mechanics. Prof. Naumann already knew the Institute for approximately 20 years. In 1937 he joined the Institute and two years later he was in charge of the gasdynamics group of the Institute. In 1946 he left the Institute of Aerodynamics to head the Bureau d'Études in Emmendingen where the new supersonic wind tunnels to be built in Vernon in France were developed.

Figure 13: Alexander Naumann.

1951 Prof. Naumann returned to Aachen to become Chief Scientist of Prof. Seewald. He headed the team that developed a supersonic wind tunnel which served as a model for larger facilities which later on were built for the DVL. From 1955 through 1966 Prof. Naumann was also Director of the DVL Institute of Applied Gasdynamics. Prof. Naumann always emphasized the link between university and institutional research. This scientific interaction was manifested by the Collaborative Research Center 83 "Fluid Mechanics and Thermo-Gasdynamics" which consisted of scientists from six university institutes and two institutes from the Deutsche Forschungs- und Versuchsanstalt für Luft- und Raumfahrt (DFVLR) which was the successor organization to the DVL. Prof. Naumann can be considered a pioneer of strong interdisciplinary research. He established the cooperation of the Department of Mechanical Engineering, the new Department of Medicine of RWTH Aachen University, and the University Hospital of Düsseldorf. It is fair to say that it was Prof. Naumann's scientific view and success that led to the foundation of the Aachen–Düsseldorf Research Group "Artificial Heart". Later on, the Collaborative Research Center 109 "Artificial Organs as Organs, Organ Substitutes, and Models" was initiated, whose Vice-Chair was Prof. Naumann, and the Helmholtz Institute for Biomedical Engineering was founded. Not only nationally but also internationally the establishment of this novel interdisciplinary research area was highly respected.

In the following, some research results that were obtained when the Institute was headed by Prof. Naumann are discussed. Detailed investigations of the stability of various vortex street configurations [45] and [46] showed that the shedding frequency was determined only by the separation angle. This result was more or less the beginning of the experimental analysis of flows over tower-like buildings and later on complete buildings such as hospitals, soccer stadiums etc. Such building aerodynamics problems became part of the DFG funded Priority Research Program "Aerodynamic Loads on Buildings". In the context of supersonic flows shock diffraction, shock decay, and the generation of secondary shocks and vortices at sharp edges with a deflection of up to 330 deg were analyzed

[47] and [48]. At low shock strength a good agreement between a pseudo–steady approximation and the experimental data was achieved. The vortex generation due to the diffraction of the shock led to the experimental investigation on shock–vortex interaction [49] and [50] using an inhouse developed camera which was able to take eight photos in the kHz range. The deformation of the shock was described by the linearized theory (Fig. 14). Furthermore, Mach–Zehnder interferograms showed the wake of a shock passing over a cuboid (Fig. 15). If the edge length in the streamwise direction was small, the vortex shed at the sharp edge became a starting vortex that was amplified by the wake of the moving shock. Even the velocity distribution of the starting vortex was captured using wolfram particles.

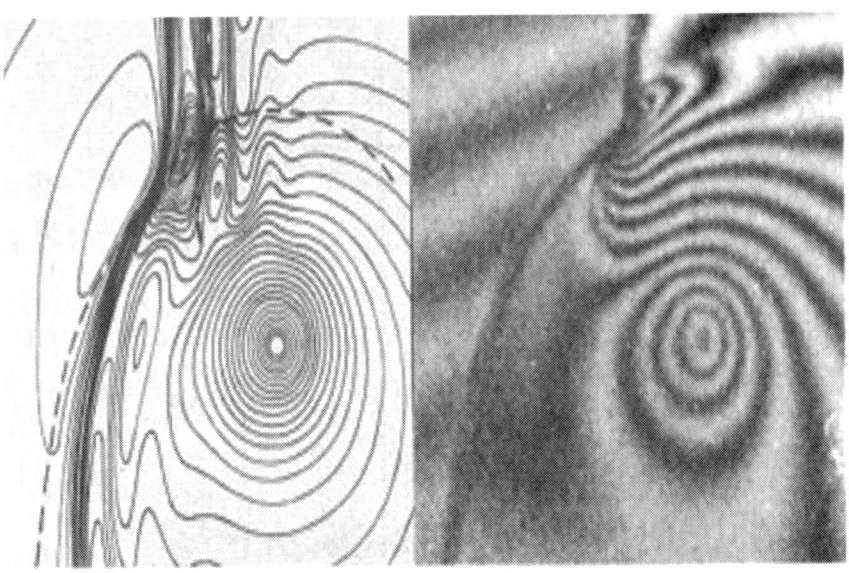

Figure 14: Shock–vortex interaction; numerical (left) and experimental (right) density contours.

Figure 15: Shock interacting with a cuboid.

From 1966 on, numerous studies were performed in biomedical flows. In [51] an analogy of unsteady gas flows in rigid pipes and incompressible unsteady flows in viscoelastic pipes was introduced. Depending on the ratio of the convective to the dissipative terms three different regimes defined the propagation speed, decay, and deformation of single waves. Non-Newtonian behavior was investigated in [52]. It was theoretically and experimentally shown that the wave velocity of a micro-polar fluid is greater than that of a Newtonian fluid. Flows through blood vessel bifurcations which are characterized by spiral-like streamlines with separation regions being prone to arteriosclerotic deposit were studied in [53] and [54]. Hot-film measurements in pulsatile flow evidenced high wall-shear stresses in the detachment region which could be the reason for pronounced irritation in those regions. Furthermore, flow visualizations in arteriovenous anastomoses which are susceptible to secondary thromboses and aneurysms were studied in [55]. The analysis of the flow field of artificial heart valves was a central research field in the studies of biomedical flows. To experimentally investigate the dynamics of the valve motion a complete blood-circuit simulator was developed [56] (Fig. 16).

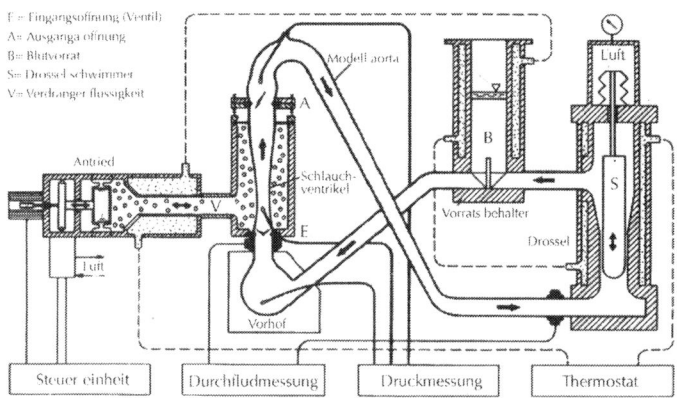

Figure 16: Artificial blood circuit to measure the blood damage rate caused by blood pumps.

For sphere prostheses, accelerations up to 60 g caused by the impact of the sphere on the rack were measured. This extreme load was conjectured to be the reason for the increased blood damage. Other valve geometries were analyzed in detail and due to an improved flow pattern the blood damage rate could be decreased leading to higher quality artificial heart valves. It was the objective of the research group "Artificial Heart" to develop a low drag and less blood damaging blood pump such that it could be used longer than standard blood pumps, i.e., for more than just a few hours. Several experimental setups were established to determine the destruction of the red blood cells caused by artificial heart valves, blood pumps, artificial kidneys, and stenoses in blood vessels. Due to the temporally extreme shear stresses on the membrane of the erythrocytes the damage of the red blood cells is higher than the natural hemolysis [57] and [58]. Measurements of the hemoglobin and the encyme G-6-PDH showed the shape of the pulse wave and the strength of the pulse to have a pronounced impact on the blood damage [59]. Deposits of blood cells in stagnation regions and the residence time in these areas were measured to understand the generation of thrombosis. In cooperation with the Department of Medicine it was evidenced that the reduction of the blood coagulation time is also related to the rate of hemolysis. Even today more in-depth investigations are necessary to understand the details of the physical–chemical mechanisms that define flow induced blood damage.

Further investigations focused on the peristaltic transport in the ureter. A theoretical model was developed in [60] and a ureter probe was developed to simultaneously measure pressure via strain gauges and velocity using NTC-elements as a function of time [61] and [62]. Other investigations dealt with the development of an artificial ureter. Some studies focused on the improvement or development of measurement techniques. Hot-wires were analyzed in [63] and the differential interferometry was studied in [64]. The development of a high-speed camera to measure supersonic flows was described in [65]. It was based on the Cranz–Schardin principle. High intensity flashes at a high damping rate were generated such that eight photos at a temporal distance of microseconds could be taken. Among other applications, the camera was used to record shock oscillations in transonic flows (Fig. 17), a phenomenon that was described by Naumann in [66]. Furthermore, methods to determine distributions of various variables in vortices were introduced [67]. To be able to investigate low-density flows in pipes and nozzles, i.e., flows in the range of Knudsen numbers $10^{-4} \leq Kn \leq 10^{-5}$ and Reynolds numbers $0.1 \leq Re \leq 2000$, a new method to determine the mass flow was introduced in [68].

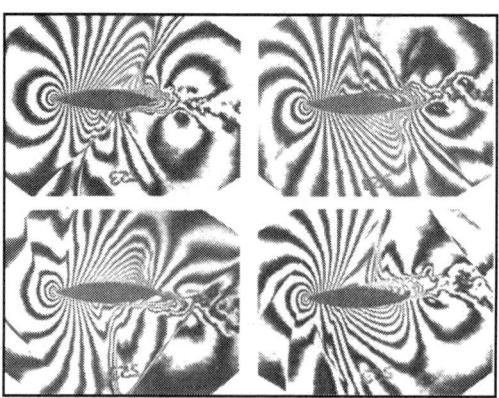

Figure 17: Flow over a circular airfoil; M=0.71, α=0°, Δt=500µs.

Since the beginning of the 70s the theoretical and experimental studies were complemented by the development of numerical integration methods for the conservation equations (Fig. 18). High-order finite-difference methods to analyze supersonic turbulent boundary layers and supersonic inviscid flows over blunt bodies were derived in [69] and [70].

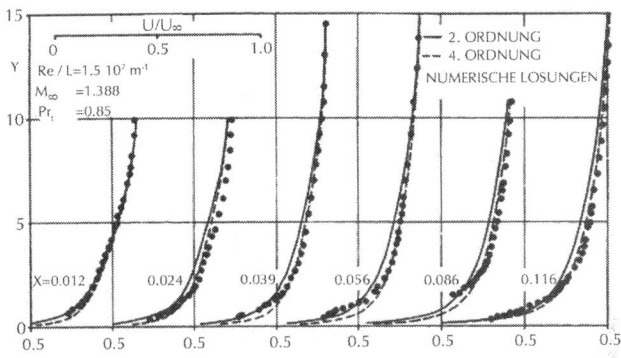

Figure 18: Comparison of numerical and experimental turbulent velocity profiles; second and fourth-order solutions.

Prof. Naumann possessed the talent to physically interpret complex flow problems which could not rigorously analyzed theoretically. He preferred the experimental investigation, and the flow visualization was his real affinity.

The period from 1973 to 1998

In 1973 Prof. E. Krause (Fig. 19) took over the Chair of Fluid Mechanics and became the 5th Director of the Institute of Aerodynamics. After his studies at the RWTH Aachen University he went to the US funded by a NATO grant to receive a Master degree in 1962 from the Polytechnic Institute in Brooklyn and a PhD degree in 1966 from the University of New York. Before he accepted the call of the RWTH Aachen University he was with the Deutsche Forschungs- und Versuchsanstalt für Luft- und Raumfahrt (DFVLR).

Figure 19: Egon Krause.

Although Prof. Krause was one of the driving factors to establish the new research field of computational fluid dynamics, the rigorous comparison of experimental and numerical results was always the basis of his scientific vision. Among other programs, he initiated the Priority Research Program "Finite Approximations in Fluid Mechanics" and the Collaborative Research Center 253 "Fundamentals of the Design of Spacecraft", which was one of three successful joint collaborative research centers in supersonic fluid mechanics funded by the DFG in the late 80s. Together with Prof. Peyret from Nice he was the pioneer of a joint long-lasting DFG–CNRS research initiative in the early 90s that brought together French and German engineers and mathematicians in the field of theoretical and computational fluid dynamics. Due to his initiative the building of the Institute of Aerodynamics was enlarged in 1978. Prof. E. Krause was awarded the Prandtl-Ring in 2004.

The studies in the field of building aerodynamics were continued throughout the Priority Research Program "Aerodynamic Loads on Buildings". Using interferometric methods and laser-Doppler anemometry the impact of a shock interacting with the wake of a flat plate on the Reynolds stress distribution was investigated [71]. Furthermore, the existence of a vortex street for a flat plate flow in the Mach number regime 1.3≤M≤2.8 was shown to depend on the roughness of the surface and the shape of the trailing edge [72] (Fig. 20).

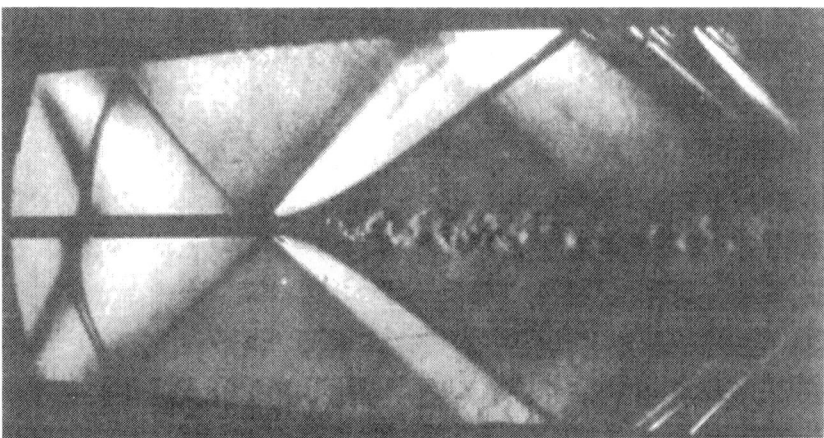

Figure 20: Vortical wake in supersonic flow at M=1.5.

In detailed Mach–Zehnder experiments of the unsteady shock–boundary-layer interaction in transonic flow Finke identified several types of interaction [73] and [74]. The possibility to influence the frequency and amplitude of the shock oscillation by suction and blowing was already discussed by Finke and later on experimentally investigated using differential interferometry in [75]. Transonic flow over supercritical airfoils and wings were experimentally and numerically analyzed in [76] and [77]. Furthermore, an airfoil with a variable upper surface contour was constructed to shift at the same freestream conditions the laminar–turbulent transition downstream due to an altered thickness distribution [78]. The impact of unsteady freestream conditions, i.e., temporally varying favorable and adverse pressure gradients, on the pressure distribution, the velocity profiles, and the wake structure of an airfoil was experimentally and numerically investigated in [79] and [80] (Fig. 21).

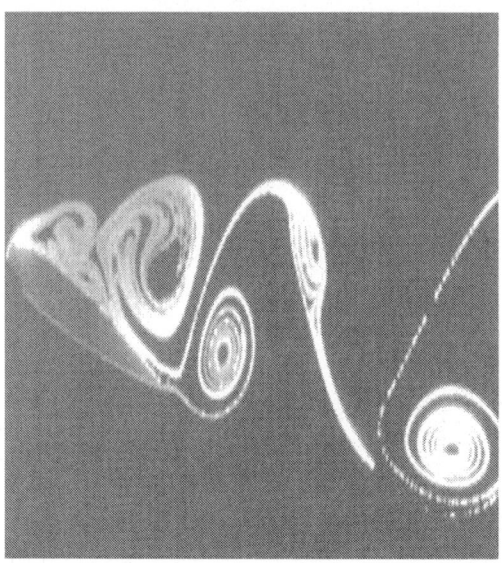

Figure 21: Numerically determined wake of a NACA0012 airfoil at =20°, Re=20000, M=0.3.

To thoroughly analyze the near-wall flow one-, two-, three-, and four-sensor hot-wire probes and the oil-film technique were further developed [81], [82] and [83]. The hot-wire probes were used to describe the time-averaged velocity profiles and the Reynolds stress distributions in a three-dimensional boundary layer [84]. The isotropy of boundary layers at very high Reynolds numbers was measured in the German–Dutch wind tunnel in [85]. Further analyses focused on perturbed boundary layers, i.e., the interaction of tangential jets [86] or the wake of a cylinder with the turbulent boundary layer were studied with respect to the Reynolds stress and energy dissipation distribution [87] and [88]. In the Priority Research Program "Transition" the multi-sensor-hot-film technique was further developed to determine transition not only in wind tunnel measurements but also at low subsonic flight tests [89] and

[90]. Stability analyses in boundary layers at supersonic freestream conditions were performed in [91] and [92]. In the 70s and 80s numerical analyses of the secondary flow in centrifuges were performed. Since the flow field and the concentration field are only weakly coupled at uranium isotope separation, the diffusion problem can be solved provided the flow field is known. Such a solution was given in [93]. Additional studies of the influence of rotation on the flow structure dealt with flows in rotating hollow spheres and Taylor vortices between two rotating concentric spheres [94] and [95]. Based on the experience in experimentally and numerically analyzing unsteady flows in safety valves, which, for instance, showed the importance of the leading edge geometry on the stability of the flow field [96] and [97], experiments and computations of the cold flow at the suction and compression strokes of a piston engine were conducted within the Collaborative Research Center 224 "Motor-Driven Combustion" [98] and [99]. The large and medium vortices of the unsteady three-dimensional flow field could be resolved at such a quality that a convincing agreement between computations and measurements was achieved [100], [101] and [102] (Fig. 22).

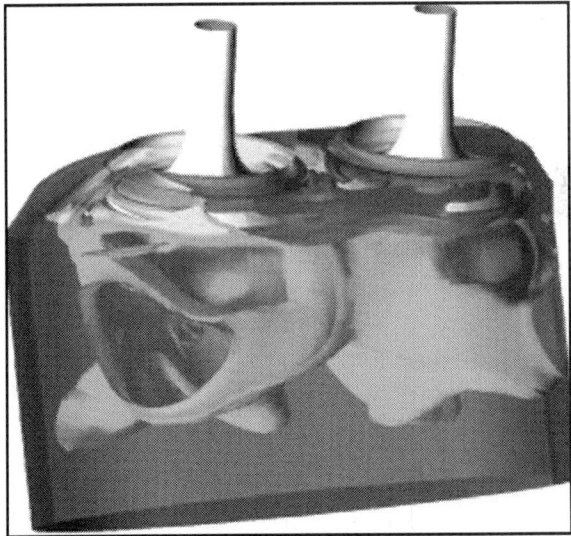

Figure 22: Vorticity contours at 150 deg crank angle.

The initialization of the Collaborative Research Center 83 "Fluid Mechanics and Thermo-Gasdynamics" marked the beginning of intensive research in the field of computational fluid dynamics. That is, numerical integration methods for the potential equations, boundary-layer equations, Euler and Navier–Stokes equations, and the Boltzmann equation were developed. Various aspects such as discretization methods, algorithms, mesh generation and so forth were considered. The quality of the numerical methods was always scrutinized by a detailed comparison of the numerical solution with experimental data. In the late 70s higher-order methods were introduced [103] and solutions for three-dimensional supersonic inviscid flows were discussed [104]. Free jets in rarefied gases were analyzed by discrete solutions of the Boltzmann equation [105] and in the context of the Collaborative Research Center 25 "Vortical Flows in Aeronautics" numerous thorough theoretical and numerical analyses of the development of slender vortices and the phenomenon of vortex breakdown were conducted [106], [107] and [108]. Detailed measurements of the three-dimensional bubble-type breakdown were discussed in [109] and spatial solutions of the temporal development bubble type (Fig. 23), spiral-type, and transitional breakdown were presented in [110] and [111].

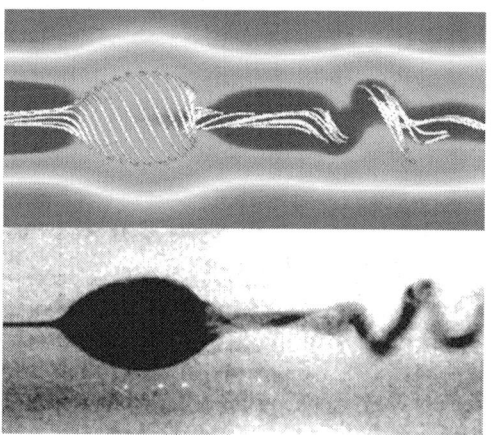

Figure 23: Bubble-type vortex breakdown; numerical vorticity lines (top) [111]; experimental streaklines (bottom) [112].

More numerical studies concerned the impact of numerical dissipation on unsteady flows [113], the computational efficiency [114] and [115], and new discretizations of the nonlinear terms of the conservation equations [116]. Algorithms for massively parallel machines were developed [117] and numerical methods to perform large-eddy simulations for internal and external flows were introduced [118] and [119]. The core of the Collaborative Research Center 253 "Fundamentals of the Design of Spacecraft" was the two-stage space-transportation system ELAC (Elliptische Auftriebskonfiguration) [120] (Fig. 24). Several numerical and experimental investigations concerning real gas effects [121], heat transfer [122], the development of primary, secondary, and tertiary vortices [123], and the scramjet propulsion concept [124] and [125]boosted supersonics research at Aachen. Further detailed experiments within the Collaborative Research Center 253 were conducted at the Institute of Theoretical and Applied Mechanics in Novosibirsk.

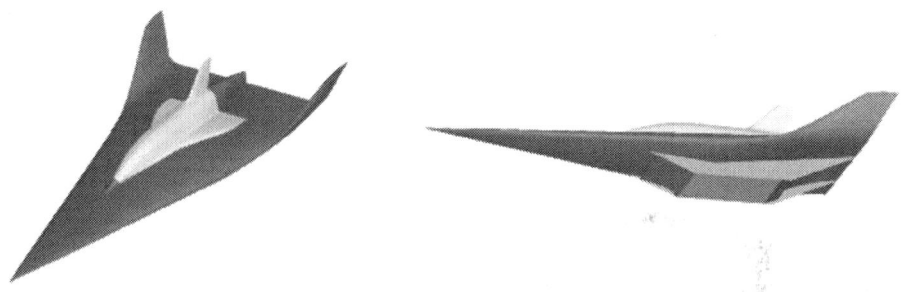

Figure 24: Two-stage space-transportation system ELAC.

The research in the field of biomedical flows also was intensified. A test facility was established to investigate the deformation of erythrocytes under increasing shear [126]. The effect of static and dynamic pressure loads on the hemolysis of human blood was published in [127]. The behavior of thrombocytes undergoing unphysiological shear loads was studied in several experiments [128] and [129]. The investigations addressed issues like short-duration shear load, extremely high shear, encyme activation, and blood coagulation. The fluid mechanics of artificial heart valves was another pronounced research field at the Institute of Aerodynamics. More or less any kind of valve type was analyzed and further developed by Köhler whose research on defining the fluid mechanical properties of artificial heart valves led to an officially certified test procedure assigned by the Department of Research and Technology [130] and [131]. Brücker extended the particle-image velocimetry technique to investigate bifurcating flows [132] and [133]. Velocity and wall-shear stress distributions in elastic vessels were measured by laser-Doppler anemometry in [134] and [135]. A test facility was designed and manufactured to investigate the flow and sound field caused by stenosis [136] and [137]. A left-heart simulator was developed (Fig. 25) that allowed among other issues the investigation of the impact of various artificial heart valve prostheses on the pressure and velocity distributions in the directly connected vessels. The comparison of numerical and experimental data showed a convincing agreement [138] and [139].

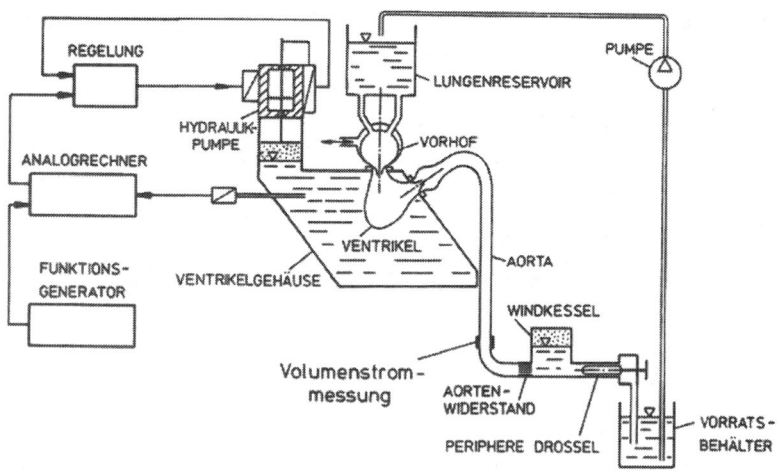

Figure 25: Schematics of a left-heart simulator.

Those investigations led to a one-dimensional model to describe the complete human blood circuit under physiologically and pathologically altered conditions [140]. Further experimental and numerical analyses considered the peristaltic transport in the ureter [141], [142] and [143]. First experimental investigations were performed for the flow field in the human nasal cavity [144], an experimental and numerical description of the bifurcating trachea flow was developed [145], and a technical model for the respiration mechanics of the lung was derived [146].

In July 1998 Prof. Krause became a professor emeritus. His scientific farewell party, which was held a couple of months later, was "sold out" within a few days. The guests who showed up from all over the world evidenced the excellent reputation he had within and across the borders of the fluid mechanics community.

THE PERIOD FROM 1998 TO 2012

In August 1998 Prof. W. Schröder (Fig. 26) accepted the call of the RWTH Aachen University, i.e., he took over the Chair of Fluid Mechanics and became the Head of the Institute of Aerodynamics. After his PhD studies at Aachen he was a postdoc at the California Institute of Technology and worked together with Prof. H. B. Keller on bifurcation problems. Then, he joined Messerschmitt-Bölkow-Blohm (MBB) where he was involved in several aerospace studies in the beginning of the 90s. Before he returned to Aachen in 1998 he held a Professorship in Mathematics and Statistics at the University of Applied Sciences in Braunschweig. Based on the long-term cooperative tradition of the Institute he strongly supports close scientific collaborations inside and outside RWTH Aachen University. The Institute participated in several Collaborative Research Centers such as the SFB 561 "Thermally Highly Loaded Porous and Cooled Multi Layer Systems for Combined Cycle Power Plants", the SFB 686 "Model Based Control of Homogenized Low-Temperature Combustion", and the SFB 401 "Modulation of Flow and Fluid-Structure Interaction at Airplane Wings" the latter of which was chaired by W. Schröder in the last funding period. Together with Profs. R. Radespiel (TU Braunschweig), N. Adams (TU Munich), and B. Weigand (Univ. of Stuttgart) he paved the way for the new Transregional Collaborative Research Center

TRR 40 in supersonic research "Fundamental Technologies for the Development of Future Space-Transport-Systems Components under High Thermal and Mechanical Loads". The interdisciplinary character of the Institute's research was strengthened by heading the Research School BrenaRo "Fuel Production from Renewable Feedstock" and chairing the DFG funded research program "Protective Artificial Respiration" which brought together scientists from medicine, physics, and engineering. Following the ideas of his predecessors W. Schröder pushes international research which is why he and Patrick Bontoux (Univ. of Marseille) launched the second phase of the French–German DFG–CNRS program. His long-term service as Treasurer for the European Mechanics Society shows his European commitment for the scientific community.

Figure 26: Wolfgang Schröder.

Supersonic research at the Institute of Aerodynamics was continued in the Collaborative Research Center 253 by analyzing the shock system at stage separation [147] and the wall-heat flux during reentry [148]. Furthermore, the bluntness transition reversal phenomenon [92] and the leading-edge receptivity in supersonic flow over a blunt flat plate was numerically investigated by spectral solutions [149]. In the Transregional Collaborative Research Center TRR 40 the dynamic base flow structures of a generic axisymmetric rocket configuration were analyzed by large-eddy simulations to understand the impact of the temporal variation on the flight characteristics and stability [150]. Supersonic vortex breakdown being caused by the interaction of slender streamwise vortices with normal and oblique shocks were experimentally investigated by differential interferometry and particle-image velocimetry as well as numerically simulated by solving the three-dimensional conservation equations of inviscid and viscous fluids [151] (Fig. 27). The issue of film cooling in supersonic flows was tackled in the transregional Graduate School GRK 1095 "Aero-Thermodynamic Design of a Scramjet Propulsion System for Future Space Transportation Systems". The large-eddy simulations (LES) clearly showed the importance of the location of the shock–cooling-film interaction on the cooling effectiveness [152]. In the Collaborative Research Center SFB 561 film cooling to protect turbine blades from high thermal loads was experimentally and numerically analyzed for numerous geometries and flow parameters. The particle-image velocimetry measurements and the large-eddy simulations showed the pronounced impact of the inflow conditions and an excellent agreement not only for the mean velocity but also for the higher-moment velocity distributions in the turbulent mixing [153] and [154].

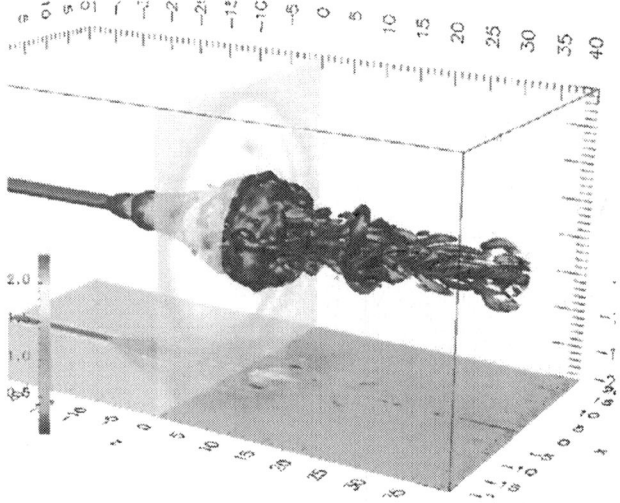

Figure 27: Vorticity and Mach number contours of normal–shock–vortex interaction.

Aerodynamic topics like the dynamics of wake vortices, laminarization, buffet, and buffeting were addressed in the Collaborative Research Center SFB 401. Using experimental and numerical methods the potential of an adaptive wing technology to improve the aerodynamic characteristics at transonic flight conditions was demonstrated in [155]. The interaction of the engine jet and the wingtip vortex in the near wake was computed and measured in [156] (Fig. 28). To better capture the free shear and vortical character that is essential in the jet–vortex interaction and in the wake a new one-equation turbulence model was derived in [157].

The influence of the position and operation mode of the engine on the wingtip vortex immediately downstream of the wing at various high-lift configurations was shown by particle-image velocimetry measurements. To better understand the physical mechanisms in unsteady transonic flow over wings, which excite dynamic instabilities such as flutter or limit-cycle oscillations, forced and free oscillations were applied at transonic flow such that the interaction of aerodynamic and structural forces on a two-dimensional rectangular wing [158] and a high aspect ratio supercritical swept wing [159] could be experimentally analyzed. Recent results confirmed the impact of the noise generated by the shear layer passing over the trailing edge on the shock oscillation [160] and [161]. The three-dimensional time-dependent flow field in the cylinder of a four-valve piston engine was measured by the particle-image velocimetry (PIV) technique and laser-Doppler anemometry the latter of which provided data to determine local energy spectra during the compression stroke [162]. Konrath also developed a holographic PIV system that was improved by Dannemann [163] in the Collaborative Research Center 686 to perform besides multiplanar monoscopic PIV the first holographic PIV measurements of the flow structure of a four-valve piston engine [164] (Fig. 29). These studies were complemented by additional stereoscopic PIV measurements and numerical analyses using a newly developed Cartesian flow solver [165] and [166]performed in the DFG funded Cluster "Taylor-Made Fuels from Biomass" and the Research School BrenaRo. Fundamental numerical and experimental investigations focused on turbulent pipe, flat plate, and channel flows. The LES based spectra of the total forces on the walls of 90 deg pipe bends showed a distinct high frequency peak due to vortex shedding and a shear layer instability. At high Reynolds numbers the spectra exhibited a low frequency oscillation caused by a fluctuation of the stagnation point at the outer side of the bend [167] (Fig. 30). The impact of spanwise oscillating surface waves on friction drag was numerically analyzed in [168]. It was evidenced that depending on the amplitude and the wave length of the surface wave the wall-shear stress could be controlled, i.e., it could be intentionally decreased and increased. The hot-wire as well as the monoscopic and stereoscopic PIV measurements for a zero-pressure gradient boundary layer at a Reynolds number based on the length of the flat plate R_{e1}=54000000 and based on the momentum thickness R_{e0}=57232, respectively, showed the turbulence in the wall-bounded shear layer to be primarily non-isotropic [169]. Turbulent channel flow was investigated by tomographic PIV measurements to confirm for the first time experimentally that dissipation elements, which were recently introduced to the turbulence literature [170], scale with the Taylor length [171]. Furthermore, PIV measurements and large-eddy simulations were performed to demonstrate the impact of rotation about the streamwise axis on turbulent channel flow [172]. In accordance with theoretical results an increased momentum exchange, i.e., a bulkier streamwise flow profile, was determined and a secondary flow perpendicular to the rotation axis developed. This fundamental knowledge was combined with the expertise in design methods and standard engineering methods to tackle more applied problems such as the development of a more efficient vertical wind turbine [173].

Figure 28: Engine-jet–wingtip-vortex interaction.

Figure 29: Holographic measurement of the flow structure in an IC engine.

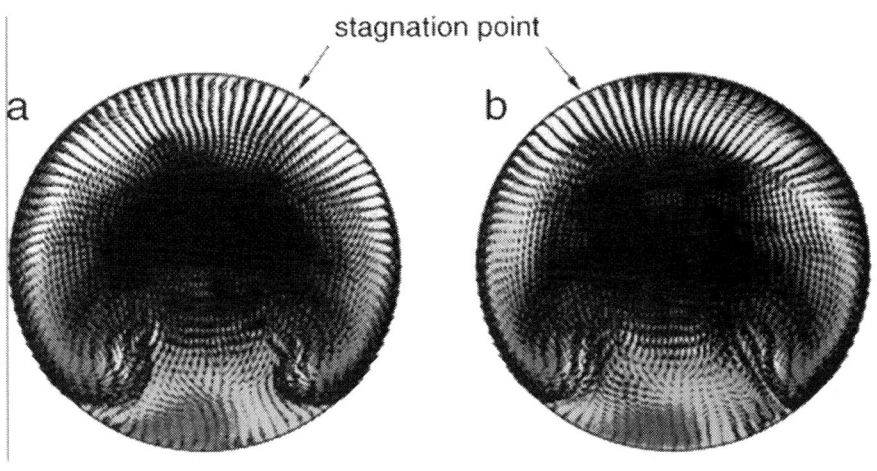

Figure 30: Temporal variation of the velocity field caused by the varying strength of the left or the right Dean vortex.

The studies on biomedical flow problems such as nasal cavity and lung flows were intensified by complementing experimental findings with numerical solutions. The simulations and the measurements for a detailed nose model evidenced that the nose flow can be well described by a laminar flow assumption [174]. Later on, a lattice–Boltzmann method was developed that allowed the efficient computation of the flow field in real nose (Fig. 31) or lung configurations [175], [176] and [177] such that together with physicians investigations for the pharmaceutical industry could be performed. Another nature inspired flow problem was the flow field of an owl at gliding flight conditions. Detailed PIV measurements concentrated on the leading-edge geometry and the velvet-like surface structure [178] to understand what makes the aerodynamics of an owl so special that hardly any aerodynamic noise is generated.

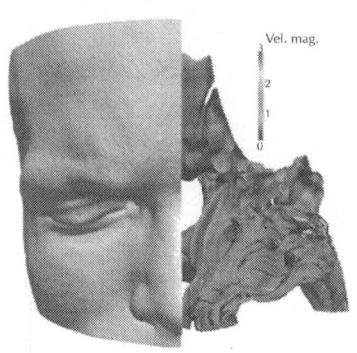
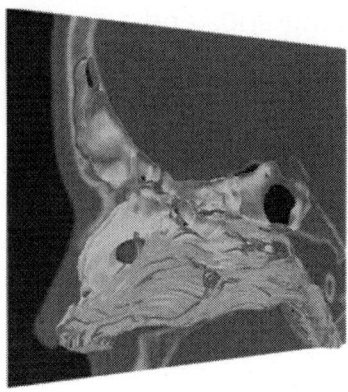

Figure 31: Streamlines of a CT-based nasal cavity flow colored by the velocity magnitude.

The analysis and the computation of flow generated noise were tackled in several theoretical and numerical investigations. A new set of acoustic perturbation equations which were analytically shown to not excite instabilities in globally unstable mean flows was derived [179] and [180] and applied to airframe noise [181] and [182] (Fig. 32) and jet noise problems [183] and [184]. Subsequently, these equations were extended to reacting flows such that also combustion generated noise can be predicted [185]. This acoustic formulation allows the analysis of the impact of the various noise sources of non-premixed and premixed flames on the overall sound field [186]. It is the objective of the acoustic research to not just analyze sound fields but to develop a technology that enables a low-noise design [187]. In the field of computational fluid dynamics several studies focused on an extension of the LES capabilities. A new inflow boundary formulation was presented in [188], a low Mach number approach was introduced in [189], and a moving grid scheme was discussed in [190]. Zonal methods, i.e., an efficient combination based on the Reynolds averaged and large-eddy formulations, were developed in [191] and [192] for subsonic flows and for transonic and supersonic problems in [193]. Furthermore, a new three-dimensional solution-adaptive Cartesian grid solver was developed including a novel cut-cell formulation that accurately accounts for embedded boundaries on non-boundary conforming grids and a feature-based mesh adaptation technique [194] and [195] (Fig. 33). The structure of the flow solver allows the analysis of geometrically highly demanding flows [166]. Besides the development of new approaches in the fields of computational aeroacoustics and computational fluid dynamics, novel ideas were introduced in the field of experimental fluid mechanics. The multi-sensor-hot-film technique was further developed in [196] such that the laminar–turbulent transition could be successfully measured at transonic free flight conditions at an altitude of approximately 10 km. The stereo-scanning particle-image velocimetry (SSPIV) [197] and the holographic PIV technique [198] were extended to investigate the three-dimensional flow of transitional separation bubbles, the vortex shedding of which is characterized in the initial stage by vortices which possess a c-like shape. To measure the time-dependent two-dimensional turbulent wall-shear stress a novel micro-pillar shear-stress sensor MPS^3 [199], [200] and [201] was developed and successfully applied to several generic problems such as pipe, duct, and flat plate flows [202]. Among other findings, the coexistence of large-scale meandering bands of low shear stress was shown (Fig. 34).

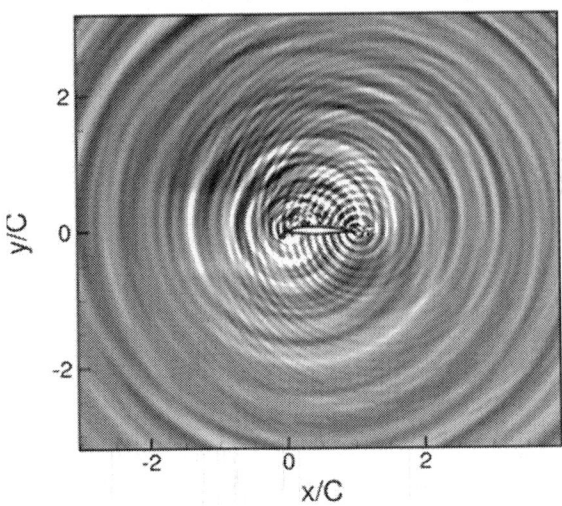

Figure 32: Acoustic pressure contours of a high-lift configuration.

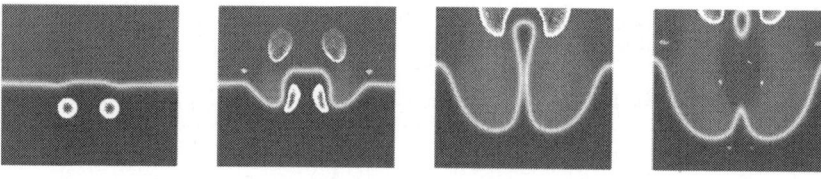

Figure 33: Flame–vortex interaction problem; temporal development of the vorticity and temperature field (from left to right).

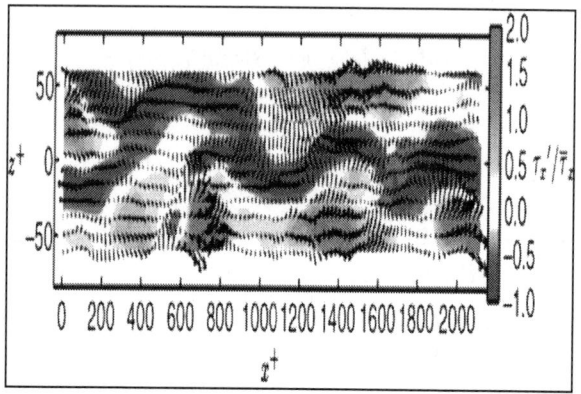

Figure 34: Wall-shear stress distribution at turbulent duct flow; contours indicate the instantaneous strength of the streamwise shear-stress fluctuations, the vectors represent the instantaneous shear-stress fluctuation.

CONCLUDING REMARKS

The research at the Institute of Aerodynamics would not have been possible without the strong support of various Departments of the State Government of Northrhine-Westfalia, the Federal Government, the German Research Association, several foundations, and highly cooperative industrial partners. It goes without saying that the research perspective of the Institute is defined by its tradition, i.e., also in the future the Institute of Aerodynamics will strive for excellence not only in fundamental research but also in being a strong link between fundamental and applied engineering science.

ACKNOWLEDGMENTS

The author would like to thank Prof. E. Krause and Drs M. Meinke and M. Klaas for sharing their knowledge on the history of the Institute of Aerodynamics that was an enormous help to write this concise article.

REFERENCES

1. A. Naumann, Aerodynamisches Institut. Abh. AIA H.17 (1963).

2. E. Krause, Das institut unter der leitung von A. Naumann, in: Abh. AIA H., vol. 22, 1975, p. 9.

3. E. Krause, Arbeiten des Instituts von 1975–1988, in: Abh. AIA H., vol. 29, 1988, p. 1.

4. E. Krause, Forschungsarbeiten des Instiuts von 1988–1998, in: Abh. AIA H., vol. 33, 1998, p. 11.

5. Th.v. Kármán, Über laminare und turbulente Reibung, ZAMM1 (1921) 233.

6. Th.v. Kármán, Über die Oberflächenreibung von Flüssigkeiten, in: Vortr. a. d. Geb. d. Hydro- und Aerodynamik, 1922, p. 146.

7. K. Pohlhausen, Zur näherungsweisen Integration der Differentialgleichung der laminaren Grenzschicht, in: Abh. AIA H., vol. 1, 1921, p. 20.

8. F. Éliás, Die Wärmeübertragung einer geheizten Platte an strömende Luft, RWTH Aachen, Dissertation, 1928,–Abh. AIA H.9 (1930).

9. Th.v. Kármán, Mechanische Ähnlichkeit und Turbulenz. Nachr. v. d. Ges. d. Wiss. zu Göttingen (1930).

10. Th.v. Kármán, Theorie des Reibungswiderstands. Konferenz über hydromechanische Probleme des Schiffsantriebs (1932).

11. L. Hopf, Der Verlauf kleiner Schwingungen auf einer Strömung reibender Flüssigkeit, Ann. d. Phys. 43 (1914) 1.

12. L. Hopf, Theorie der Turbulenz, Ann. d. Phys. 59 (1919) 538.

13. Th.v. Kármán, Über die Stabilität der Laminarströmung u. die Theorie der Turbulenz, in: Proc. of the Intern. Congress f. Appl. Mechanics, vol. 27, 1924, Abh. AIA H.4 (1925).

14. Th.v. Kármán, E. Trefftz, Potentialströmung um gegebene Tragflächenquerschnitte, ZFM 9 (1918) 111.

15. Th.v. Kármán, Th.v. Bienen, Zur Theorie der Luftschrauben, Z. VDI 68 (1924) 1237.

16. Th. Bienen, Die günstige Schubverteilung für die Luftschraube bei Berücksichtigung des Profilwiderstands, ZFM 16 (1925) 209–221.

17. Th. Troller, Aerodynamische Theorie und Entwurf von Luftschrauben, RWTH Aachen, Dissertation, 1932.

18. Th. Troller, Zur Wirbeltheorie der Luftschraube, ZAMM 8 (1928) 426.

19. Th. Troller, Zur Berechnung von Schraubenventilatoren, in: Abh. AIA H., vol. 10, 1931, p. 43.

20. K. Friedrichs, Th.v. Kármán, Zur Berechnung freitragender Flügel, Abh. AIA H.9, 1929, 1-ZAMM9 (1929), 261.

21. C. Wieselsberger, Die aerodynamische Waage des Aachener Windkanals, in: Abh. AIA H., vol. 14, 1934, p. 24.

22. C. Wieselsberger, Über die Verteilung des Auftriebs längs der Spannweite bei hohen Anstellwinkeln. FB 493 (1935).

23. C. Wieselsberger, On the Distribution of Lift Across the Span Near and Beyond the Stall. Journ. Aeron. Sci. New York, 4 (1936/1937) 363.

24. C. Wieselsberger, Die Überschallanlage des Aerodynamischen Instituts der Techn. Hochschule Aachen, in: Luftwissen, vol. 4, 1931, p. 301.

25. C. Wieselsberger, Disskussionsbem. in: Convegno Di Scienze Fisiche, Mat. e. Nat. Roma., 1935, p. 558.

26. R. Hermann, Der Kondensationsstoß in Überschall–Windkanal-Düsen, LuFo 19 (1942), 201.

27. C. Wieselsberger, Elektrische Anzeige von Kräften durch Änderung einer Induktivität. Jb. 1937 LuFo, in: Ausg. Flugwerk SI, vol. 592, 1937, p. 3.

28. C. Wieselsberger, Über den Einfluß der Windkanalbegrenzung auf den Widerstand im Bereich der kompressiblen Strömung. FB 1172 (1939).

29. F. Bollenrath, J. Nemes, Über das Verhalten verschiedener Leichtmetalle in der Kälte, Metallwirtschaft 10 (1931) 609–615.

30. H. Eggink, Strömungsaufbau und Druckrückgewinn in ÜberschallWindkanälen, RWTH Aachen, Dissertation, 1943,–FB 1756 (1943).

31. W. Linke, Die Bemessung von Kühlstoffkühlern. FB 1678 (1942).

32. W. Linke, Kühlersysteme und Kühlereinbau für den Höhenflug. Ber. 161 Lil. Ges. (1943).

33. P. Hadlatsch, Einfluß von Gasschwingungen endlicher Amplitude auf den Spüvorgang eines Zweitaktmotors, RWTH Aachen, Dissertation, 1948.

34. H. Zeller, Beitrag zur eindimensionalen stationäre Gasströmung mit Reibung und Wärmeleitung, insbesondere in Rohren mit unstetigen Querschnittsänderungen, RWTH Aachen, Dissertation, 1954.

35. W. Wilhelm, Untersuchung über den Einfluß der Auspuffrohrabmessung auf den Ladungswechsel einer Einzylinder-Zweitakt-Vergasermaschine mit Kurbelkastenspülung. FB 588 NRW (1958).

36. W. Wilhelm, Die Wirkung von Auspuffrohren mit Blenden am Rohrende sowie diffusorartiger Auspuffleitungen auf den Ladungswechsel einer Einzylinder-Zweitakt-Vergasermaschine mit Kurbelkastenspülung. FB 982 NRW (1961).

37. H.F. Klein, Die nichtstationären Strömungsvorgänge und der Wäremübergang in einem Schwingfeuergerät, RWTH Aachen, Dissertation, 1956.

38. P. Hadlatsch, Berechnung der Druckwellen in Brennstoffeinspritzsystemen und in hydraulischen Ventilsteuerungen. FB 981 NRW (1961).

39. J.A. Khan, Untersuchung zur instationären Strömung durch unstetige Querschnittsänderungen in Druckleitungen von Einspritzsystemen, RWTH Aachen, Dissertation, 1960.

40. H. Rosenberg, Instationäre Strömungsvorgänge in Leitungssystemen mit flexibel-elastischen Rohrwänden, RWTH Aachen, Dissertation, 1963.

41. A. Naumann, Versuche an aerodynamischen Drosselsystemen, ZFW 9 (1961) 117.

42. E. Hübner, Wärmeübergang bei pulsierend strömendem Öl, RWTH Aachen, Dissertation, 1952.

43. H. Hasselgruber, Die Berechnung der Temperaturen in Reibungskupplungen, RWTH Aachen, Dissertation, 1953.

44. W. Linke, Strömungsvorgänge in zwangsbelüfteten Räumen, VDI Ber. 21 (1957) 29.

45. A. Naumann, M. Morschbach, C. Kramer, The conditions of separation and vortex formation past cylinders, in: AGARD Conference Pro. 4, Separated Flows 2 (1966) 539 574.

46. A. Naumann, H. Quadflieg, Aerodynamic aspects of wind effects on cylindrical buildings, in: Proc. Sympos. Wind Effects on Buildings and Structures 1, Loughborough (1968)–Pap. 9.

47. S. Schultz, Über die Beugung von Stoßwellen an scharfen Kanten, in: Abh. AIA H., vol. 20, 1970, p. 31.

48. S. Schultz, Über die Beugung von Stoßwellen an scharfen Kanten, ZFW Bd.20 (1972) 179.

49. E. Hermanns, Die Ausbreitung von ebenen Stoßfronten in dem Strömungsfeld eines einzelnen Wirbels, RWTH Aachen, Dissertation, 1972.

50. A. Naumann, E. Hermanns, On the interaction between a shock wave and a vortex field, in: AGARD Conf. Proc. 131 (1973)–Pap. 23.

51. J. Lorenz, H. Zeller, An Analogons Treatment of Wave Propagation in Liquidfilled Elastic Tubes and Gas-filled Rigid Tubes. Proc. 1th Int. Conf. on Pressure Surges, Canterbury (1972).

52. R. Ravindran, Propagation of pressure pulse in non-Newtonian fluids a theoretical and experimental investigation, J. Biorheology 11 (1974) 197–205.

53. H. Zeller, N. Talukder, J. Lorenz, Model studies of pulsating flow in arterial branches and wave propagation in blood vessels, in: AGARD Conference Proc., vol. 65, 1970.

54. N. Talukder, H. Zeller, Modellversuche zur Strömung in arteriellen Verzweigungen, in: DGBMT Hannover, 1974, pp. 163–164.

55. S. Lymberopoulos, H. Melchior, R. Gerlach, Die End-zu-End innere AVAnastomose zur Hämodialyse, Verhandlungsbericht der Gesellschaft für Urologie (1974) 326–330.

56. J. Köhler, C. Kramer, Zum Schließvorgang künstlicher Herzklappen, Acta Medictechnica Bd. 20 (1972) 36–41.

57. A. Naumann, C. Kramer, Flow investigations on artificial heart valves, AGARD Conference Proc. 65 (1970).–Pap. 4.

58. C. Kramer, Studies on flow induced mechanical Haemolysis. AGARD Conference Proc. 65 (1970)–Pap. 5.

59. J. Lambert, M. Egberts, H. Holzhüter, E. Wenzel, N. Faude, Experimentelle Ermittlung der zeitlichen Druckänderung und Blutschädigung im Eingriffsbereich einer Rollenpumpe. DGBMT, Hannover (1974) 173.

60. K.K. Simhan, Über ein theoretisches Modell zur Erfassung des peristaltischen Transportvorganges. Melchior-Lutzeyer, Urodynamisches Symposium, 1971, pp. 60–69.

61. H. Melchior, K.K. Simhan, P. Rattert, W. Lutzeyer, Simultaneuos Pressure and Flow Monitory in the Ureter. Symp. on Flow-measurement, Pittsburgh (1971)–Pap. 4-5-228.

62. H. Melchior, K.K. Simhan, A new Uro-Rheomanometer, Urodynamics (1973) 30

63.   H. Schollmeyer, Vergleich der theoretisch ermittelten und der gemessenen Temperaturen eines Hitzdrahtes endlicher Länge, in: DLR Mitt., vol. 64-04, 1964, pp. 65 77.

64.   C. Kramer, Die Differentiallinterferometrie als Meßverfahren der gasdynamischen Forschungsarbeiten, in: Abh. AIA H., vol. 18, 1965, p. 37.

65.   E. Hermanns, C. Kramer, M. Meszerits, S. Schultz, Entwicklung einer Mehrfunken Kamera für Strömungsuntersuchungen, in: Abh. AIA H., vol. 20, 1970, p. 15.

66.   A. Naumann, Stoßschwingung an Profilen, in: Abh. AIA H., vol. 18, 1965, p. 9.

67.   E. Hermanns, Methoden zur Bestimmung der Zustandsgrößen in realen Wirbeln, in: Abh. AIA H., vol. 20, 1970, p. 20.

68.   A. Naumann, C.H. Chun, Mass flow measurements for sharp-edged orifices in low density flows, in: IX th Int. Symp. on Rarefied Gas Dyn. (1974)–Pap. D 21.

69.   E. Krause, E.H. Hirschel, W. Kordulla, Fourth order "Mehrstellen"—integration for three-dimensional turbulent boundary layers, in: AIAA Computational Fluid Dynamics Conf. Proc., 1973, p. 92.

70.   C. Weiland, Berechnung dreidimensionaler Umströmungen stumpfer Körper im Überschall mit dem Rusanow-Verfahren, in: Abh. AIA H., vol. 21, 1974, pp. 58–61.

71.   G. Marenbach, Störung ebener, turbulenter Mischzonen durch Stoßwellen, RWTH Aachen, Dissertation, 1982.

72.   M. Tang, Wirbelstraße im Überschallnachlauf einer rauhen, ebene Platte, RWTH Aachen, Dissertation, 1983.

73.   K. Finke, Unsteady shock-wave boundary-layer interaction on profiles in transonic flows, in: AGARD Conference Proc., vol. 168, 1975, p.28.

74.   K. Finke, Stoßschwingungen in schallnahen Strömungen, RWTH Aachen, Dissertation, 1977.

75.   T. Franke, Experimentelle Untersuchung schallnah angeströmter Tragflügelprofile in Kanälen mit geringer Meßkammerbreite, RWTH Aachen, Dissertation, 1987.

76. H.-J. Romberg, Experimentelle Untersuchung der schallnahen Umströmung eines superkritischen Tragflügels unter besonderer Berücksichtigung von Windkanalinterferenzen, RWTH Aachen, Dis- sertation, 1990.

77. G. Seider, D. Hänel, Numerical Influence of Upwind TVD Schemes on Transonic Airfoil Drag Prediction. AIAA Paper 91-0184 1991.

78. P. Guntermann, G. Dietz, Laminarexperimente mit verstellbarem Flügelprofil, Luft- und Raumfahrt, 1992, Heft 6.

79. E. Krause, G. Ehrhardt, B. Schweitzer, Experiments on unsteady flows about wing sections. Proceedings of the Conference on Low Reynolds Number Airfoil Aerodynamics. UNDAS-CP-77B123.

80. K. Dortmann, Computation of viscous unsteady compressible flow about airfoils, Proc. of the 11th ICNMFD,–Springer Verlag.

81. U.R. Müller, On the accuracy of turbulence measurements with inclined hot wires, J. Fluid Mech. 119 (1982) 153–155.

82. T. Schön, U.R. Müller, A new hot-wire technique for measuring the instantaneous velocity vector in highly turbulent flows, in: Second European Turbulence Conference, 1988.

83. U.R. Müller, F. Feyzi, Skin-friction measurements by laser beam interferometry, AIAA J. (1989) 984.

84. U.R. Müller, Measurement of the Reynolds stresses and the mean-flow field in a three-dimensional pressure-driven boundary layers, J. Fluid Mech. 119 (1982) 121–152.

85. H.H. Fernholz, E. Krause, M. Nockemann, M. Schober, Comparative measurements in the canonical boundary layer at $Red2£ = 6 \times 104$ on the wall of the German–Dutch windtunnel, Phys. Fluids 7 (1995) 1275–1281.

86. N.I.I. Hewedy, Untersuchung eines ebenen, tangential ausgeblasenen Gegenstroms, RWTH Aachen, Dissertation, 1980.

87. U.R. Müller, J. Wu, Experimental investigation of turbulence energy dissipation rate in a relaxing boundary layers, in: Proc. 6th Symposium on Turbulent Shear Flows, Toulouse, 1987.

88. U.R. Müller, T. Schön, Developments towards a three-component hot-wire technique for high intensity and separated turbulent flows, in: Proc. 6th Symposium on Turbulent Shear Flows, Toulouse, 1987.

89. M. Kronberger, Multisensor-Heißfilmtechnik zur Transitionserkennung im Windkanal- und Flugversuch, RWTH Aachen, Dissertation, 1992.

90. St. Fühling, Untersuchung transitioneller, ablösender und gestörter Grenzschichten mit der Multisensor-Heißfilmtechnik, RWTH Aachen, Dissertation, 1999.

91. G. Dietz, A. Meijering, Numerical investigation of the boundary layer instabilities over a blunt flat plate at angle of attack in supersonic flows, in: Notes on Numerical Methods, vol. 60, 1997, pp. 103–110.

92. G. Dietz, Entropie- und Grenzschichtinstabilitäten an einer stumpfen ebenen Platte in Überschallströmung, RWTH Aachen, Dissertation, 1999.

93. A. Merten, D. Hänel, Diffusion and flow calculations in centrifuges using the full conversation equations. Proc. 5, Workshop on Gases in Strong Rotations, Charlotteville (1984).

94. F. Bartels, Taylor vortices between two concentric rotating spheres, J. Fluid Mech. 119 (1982) 1–25.

95. G. Schrauf, E. Krause, Symmetric and asymmetric Taylor vortices in a spherical gap, in: IUTAM Symposium Novosibirsk 1985.

96. B. Föllmer, H. Zeller, The influence of pressure surges on the functioning of safety valves. Proc. Internat. Conf. on Pressure Surges, 3–Bedford/U.K, pap. J2.

97. [97] U. Giese, Berechnung schallnaher Einlaufströmungen, RWTH Aachen, Dissertation, 1983.

98. H. Henke, D. Hänel, Numerical simulation of gas motion in piston engines, in: Proc. of the Ninth Intern. Conf. on Num. Methods in Fluid Mech., 1984, 267–271.

99. B. Binninger, Untersuchung von Modellströmungen in Zylindern von Kolbenmotoren, RWTH Aachen, Dissertation, 1989.

100. J. Klöker, Berechnung von Wirbelstrukturen im Zylindern eines Modellmotors, in: Abh. AIA H., vol. 31, 1994, p. 1.

101. R. Ortmann, Untersuchung der Strömung im Zylinder eines 4-ZylinderSerienmotors mit der Particle-Image-Velocimetry, RWTH Aachen, Dissertation, 1997.

102. A. Abdelfattah, Numerische Simulation von Strömungen in 4-VentilMotoren, RWTH Aachen, Dissertation, 1998.

103. E. Krause, E.H. Hirschel, W. Kordulla, Fourth order "Mehrstellen"—integration for three-dimensional turbulent boundary layers, Comput. Fluids 4 (1976) 77–92.

104. C. Weiland, Lösung der Euler-Gleichung für räumlich Überschallströmungen um stumpfe Körper, RWTH Aachen, Dissertation, 1975.

105. D. Bergers, Kinetic model solution for axisymmetric flow by method of discrete ordinates, J. Comput. Phys. 57 (1985) 285–302.

106. P.-M. Hartwich, Berechnung von Vorderkantenwirbeln an Deltaflügeln, RWTH Aachen, Dissertation, 1983.

107. E. Krause, A contribution to the problem of vortex breakdown, Comput. Fluids 13 (1985) 375–381.

108. L. Reyna, S. Menne, Numerical prediction of flow in slender vortices, Comput. Fluids 16 (3) (1988) 239–256.

109. Ch. Brücker, W. Althaus, Study of vortex breakdown by particle tracking velocimetry (PTV) part 1: bubble-type vortex breakdown, Exp. Fluids 13 (1992) 339–349.

110. M. Breuer, Numerische Simulation des Wirbelaufplatzens, Abh. AIA H.32, 1994, pp. 44–57.

111. M. Weimer, Aufplatzen freier Wirbel und drallbehafteter Rohrströmungen, RWTH Aachen, Dissertation, 1997.

112. J.H. Fahler, S. Leibovich, Disrupted states of vortex flow and vortex breakdown, Phys. Fluids 20 (1977) 1385–1400.

113. H. Henke, D. Hänel, Artificial damping in approximate factorization methods in: Proc. of 6th GAMM Conference on Numerical Methods in Fluid Mechanics, 1986, pp. 137–145.

114. W. Schröder, D. Hänel, An unfactored implicit scheme with multigrain acceleration for the solution of the Navier–Stokes equations, Comput. Fluids 15 (3) (1987) 313–336.

115. M. Meinke, Numerische Lösung der Navier–Stokes-Gleichungen für instationäre Strömung mit Hilfe der Mehrgittermethode, RWTH Aachen, Dissertation, 1993.

116. R. Schwane, D. Hänel, An implicit flux-vector splitting scheme for the computation of viscous hypersonic flows. AIAA Paper 89-0274 1989.

117. J. Hofhaus, D. Van, E.F. Velde, Alternating-direction line-relaxation methods on multicomputers, SIAM J. Sci. Compt. 17 (2) (1996) 454–478.

118. C. Schulz, Grobstruktursimulation der Mischungsprozesse in turbuleneten Freistrahlen, RWTH Aachen, Dissertation, 1997.

119. T.K. Rister, Grobstruktursimulation schwach kompressibler turbulenter Freistahlen-ein Vergleich zweier Lösungsansätze, RWTH Aachen, Dissertation, 1998.

120. E. Krause, German University Research in Hypersonics. AIAA Paper 92-5033 (1992).

121. A. Henze, Integration der Navier–Stokes-Gleichungen für Hyperschallströ- mungen mit Realgaseffekten, RWTH Aachen, Dissertation, 1995.

122. M. Jacobs, Wärmeübergangsmessung am Modell eines Raumtrans- portsystems in supersonischer Strömungs, RWTH Aachen, Dissertation, 1997.

123. N. Lang, Sichtbarmachung und Geschwindigkeitsmessung in Überschallleeseitenwirbel mit Lichtschnittverfahren, RWTH Aachen, Dissertation, 2000.

124. M. Kropp, Reagierende Über- und Hyperschallumströmung eines Raumtransportsystems mit Außenverbrennung, RWTH Aachen, Dissertation, 1998.

125. E. Schmitz, Kalte Triebwerkssimulation am Raumtransportsystem ELAC 1, RWTH Aachen, Dissertation, 2002.

126. A. Naumann, J. Lambert, Das Verhalten roter Blutkörperchen unter der Einwirkung kurzzeitiger, hoher laminarer Schubspannungen, Biomed. Tech. Ergänzungsbd. 21 (1976) 75–76.

127. R. Gerlach, J. Lambert, Auswirkung statischer und dynamischer Druckbelastung auf die Hämolyserate von Humanblut, Biomed. Tech. Ergänzungsbd. (1975) 65–66.

128. G. Heuser, Secondary effects in cone and plate viscometers, Biorheology 15 (1978) 311–320.

129. G. Heuser, R. Opitz, A coquette viscometer for short time shearing of blood, Biorheology 17 (1980) 17–24.

130. J. Köhler, Prosthetic heart valves and similarity, Med. & Biol. Eng. 23 (1985) 649–650.

131. J. Köhler, Quality assurance of artificial heart valves, in: Vitro Testing, Heart Valve Engineering, Mechanical Engineering Publications Ltd., 1986, pp. 89–98.

132. Ch. Brücker, 3-D Scanning PIV applied to an air flow in a motored engine using digital high-speed video, Meas. Sci. Technol. 8 (1997) 1480–1492.

133. Ch. Brücker, M.L. Riethmüller, Cyclic flow oscillation in a system of repeatedly branching channels, Phys. Fluids 10 (4) (1998) 877–885.

134. D. Tomm, Bestimmung der Geschwindgkeitsverteilung in Gefäßverzweigungsmodellen mittels Laser-Doppler-Anemometer, Biomed. Tech. Ergänzungsbd. (1975) 17.

135. D. Tomm, Determination of steady and unsteady wall shear from LaserDoppler anemometer measurements. Proc. Euromech 90 1977.

136. D. Tomm, Model investigation of noise generation in vessel stenoses, Cardiovascular Pulmonary Dynamics, EUROMECH 92 (1978) 179–192.

137. J. Reinecke, Strömung und Geräusch in Modellen arterieller Stenosen, RWTH Aachen, Dissertation, 1982.

138. B. Steinbach, W. Limberg, R. Optiz, W. Bialonski, Simulation of the interaction heart-pump to improve assist-techniques with minimal myocardial lesions, Biomed. Tech. 28 (1983) 235–241.

139. W. Bialonski, Modellstudie zur Entlastung des linken Herzens, RWTH Aachen, Dissertation, 1987.

140. M. Zacek, E. Krause, Numerical simulation of the blood flow in the humen cardiovascular systems, J. Biomech. 29 (1) (1996) 13–20.

141. R. Gerlach, S. Lymberopoulos, A new method for artificial urinary conveyance in the upper urinary tract, Proc. Eur. Soc. for Artif. Organs 2 (1976) 179.

142. R. Gerlach, M. Graw, Untersuchung der Strömung im Harnleiter und künstlicher Ersatz des Organs, Abh. AIA H. 25, 1980, pp. 48–49.

143. H. Grave, Numerische und experimentelle Simulation peristaltischen Transports, RWTH Aachen, Dissertation, 1987.

144. R. Opitz, W. Limberg, Experimentelle Untersuchung der Strömung in einem Nasenmodell, in: Abh. AIA H., vol. 32, 1996, p. 120.

145. R. Teichmann, Alternierende Verzweigungsströmungen am Beispiel des bronchialen Systems, RWTH Aachen, Dissertation, 1993.

146. W. Limberg, R. Opitz, L. Lorang, Theoretical and experimental studies based on a new model of the mechanics of breathing. SEB Symposium Biological Fluid Dynamics, University of Leeds (1994).

147. C. Brodbeck, Entwicklung eines strukturiert/unstrukturierten Verfahrens zur Lösung der Navier–Stokes Gleichungen, RWTH Aachen, Dissertation, 2003.

148. C.-C. Ting, Strömungs- und Wärmeübergangsmessung für das zweistufige Raumtransportsystem ELAC, RWTH Aachen, Dissertation, 2003.

149. St. Mählmann, W. Schröder, Spectral simulation of disturbance evolustion in the leading-edge region of blunt flat plate in supersonic flow, Fluid Dyn. Res. 49 (2008) 803–826.

150. J.-H. Meiß, Numerical Investigation of Nozzle-Base Flow of a Generic Space Vehicle, RWTH Aachen, Dissertation, 2009.

151. M. Klaas, O. Thomer, W. Schröder, Experimental and Computational Investigation of Oblique Shock-Vortex Interaction. AIAA Paper 2002-3305 (2002).

152. M. Konopka, M. Meinke, W. Schröder, Large-eddy simulation of shockcooling-film interaction, AIAA J. 50 (10) (2012) 2102–2114.

153. P. Renze, W. Schröder, M. Meinke, LES of turbulent mixing in film cooling flows, Flow, Turbulence and Combustion 80 (2008) 119–132.

154. W. Jessen, M. Konopka, W. Schröder, Particle-image velocimetry measurements of film cooling in an adverse pressure gradient flow, in: ASME Journal of Turbomachinery, vol. 2, 2010, p. 134.

155. A. Meijering, Design of Adaptive Wind Sections with Natural Transition, RWTH Aachen, Dissertation, 2003.

156. E. Fares, G. Huppertz, R. Abstiens, W. Schröder, Investigation of the engine jet/wind tip vortex interference, Aerosp. Sci. Technol. 8/3 (2003) 175–183.

157. E. Fares, W. Schröder, A general one-equation turbulence model for free shear and wall-bounded flows, Flow, Turbul. Combust. 73 (2004) 187–215.

158. H.C. Hillenherms, Experimental Investigation of a Super-Critical Airfoil Oscillating in Pitch at Transonic Flow, RWTH Aachen, Dissertation, 2003.

159. P.Ch. Steimle, D. Karhoff,W. Schröder, Unsteady transonic flow at a transporttype wing, AIAA J. 50 (2) (2012) 399–415.

160. A. Hartmann, P.Ch. Steimle, M. Klaas, W. Schröder, Time resolved particleimage velocimetry of unsteady shock-boundary-layer interaction, AIAA J. 49 (1) (2011) 195–204.

161. A. Hartmann, M. Klaas, W. Schröder, Time resolved stereo PIV measurements of shock-boundary-layer interaction on a supercritical airfoil, Exp. Fluids 52 (2012) 591–604.

162. R. Konrath, Strömungsanalyse im Zylinder eines 4-Ventil-Motors mittels der holographischen Particle-Image Velocimetry, RWTH Aachen, Dissertation, 2003.

163. J. Dannemann, K. Pielhop, M. Klaas, W. Schröder, Cycle-resolved multi-planar flow measurements in a four-valve combustion engine, Exp. Fluids 50 (2011) 961–976.

164. J. Dannemann, Multidimensional Measurements of the In-Cylinder Flow of a Four-Valve Internal Combustion Engine, RWTH Aachen, Dissertation, 2012.

165. D. Hartmann, M. Meinke, W. Schröder, An adaptive multilevel formulation for Cartesian hierarchical grid methods with adaptive mesh refinement, Comput. Fluids 37 (2008) 1103–1125.

166. C. Günther, L. Schneiders, M. Meinke, W. Schröder, D. Hartmann, A Cartesian cut-cell method for sharp moving boundaries. AIAA Paper 2011-3387 (2011).

167. F. Rütten, W. Schröder, M. Meinke, LES of low frequency oscillations of the dean vortices in turbulent pipe bend flows, Phys. Fluids 17 (2) (2005) 035107.

168. S. Klumpp, M. Meinke, W. Schröder, Friction drag variation by spanwise transversal surface waves. Flow, Turbulence and Combustion (2011). http://dx.doi.org/10.1007/s10494-011-9326-3 (published online).

169. R. Abstiens, Experimentelle Untersuchung der turbulenten Strukturen in einer Plattengrenzschicht bei groen Reynoldszahlen, RWTH Aachen, Dissertation, 2007.

170. L. Wang, N. Peters, The length-scale distribution function of the distance between extremal points on passive scalar turbulence, J. Fluid Mech. 554 (2006) 457–475.

171. L. Schäfer, U. Dierksheide, M. Klaas, W. Schröder, Investigation of dissipation elements in a fully developed turbulent channel flow by tomographic particle-image velocimetry, Phys. Fluids (2011) 35–106.

172. I. Recktenwald, N. Alkishriwi, W. Schröder, PIV-LES analysis of channel flow rotating about the streamwise axis, Eur. J. Mech. B/Fluids 28 (2009) 677–688.

173. M. Marnett, Multiobjective Numerical Design of Vertical Axis Wind Turbine Components, RWTH Aachen, Dissertation, 2012.

174. I. Hörschler, Ch. Brücker, W. Schröder, M. Meinke, Investigation of the impact of the geometry on the nose flows, Eur. J. Mech. B/Fluids 25 (4) (2006) 471–490.

175. R.K. Freitas, Analysis of Lattice–Boltzmann Methods for Internal Flows, RWTH Aachen, Dissertation, 2008.

176. R.K. Freitas, W. Schröder, Numerical investigation of the three-dimensional flow in a human lung model, J. Biomech. 41 (2008) 2446–2457.

177. Th. Soodt, F. Schröder, M. Klaas, T. van Overbrüggen, W. Schröder, Experimental investigation of the transitional bronchial velocity distribution using stereo scanning PIV, Exp. Fluids 52 (2012) 709–718.

178. St. Klän, S. Burgmann, Th.v. Bachmann, M. Klaas, H. Wagner, W. Schröder, Surface structure and dimensional effects on the aerodyanmics of an owlbased wing model, Eur. J. Mech. B/Fluids.

179. R. Ewert, A Hybrid Computational Aeroacoustics Method to Simulate Airframe Noise, RWTH Aachen, Dissertation, 2002.

180. R. Ewert, W. Schröder, Acoustic perturbation equations based on flow decomposition via source filtering, J. Comput. Phys. 188 (2003) 365–398.

181. R. Ewert, W. Schröder, On the simulation of trailing edge noise with a hybrid LES/APE methods, J. Sound Vib. 270 (2004) 509–524.

182. D. König, S.R. Koh, M. Meinke, W. Schröder, Two-step simulation of slat noise, Comput. Fluids 39 (2010) 512–524.

183. E. Gröschel, W. Schröder, P. Renze, M. Meinke, P. Comte, Noise prediction for a turbulent jet using different hybrid methods, Comput. Fluids 37 (4) (2008) 414–426.

184. S.R. Koh, W. Schröder, M. Meinke, Turbulence and heat excited noise sources in single coaxial jets, J. Sound Vib. 329 (2010) 786–803.

185. T.Ph. Bui, Theoretical and Numerical Analysis of Broadband Combustion Noise, RWTH Aachen, Dissertation, 2008.

186. T.Ph. Bui, W. Schröder, M. Meinke, Acoustic perturbation equations for reacting flows to compute combistion noise, Int. J. Aeroacoustics 6 (4) (2007) 335–355.

187. S. Koh, M. Meinke, W. Schröder, Airframe-noise reduction by suppressing near-wall turbulent structures. AIAA Paper 2011-2904 (2011).

188. W.A. El-Askary, Zonal Large Eddy Simulations of Compressible Wall-Bounded Flows, RWTH Aachen, Dissertation, 2004.

189. N. Alkishriwi, M. Meinke, W. Schröder, A large–eddy simulation method for low mach number flows using preconditioning and multigrid, Comput. Fluids 35 (10) (2006) 1126–1136.

190. M. Opiela, Grobstruktur-Simulation der Interaktion des Nachlaufs eines bewegten Kreiszylinders mit einer Turbinenschaufel, RWTH Aachen, Dissertation, 2003.

191. Q. Zhang, W. Schröder, M. Meinke, A zonal RANS/LES method to determine the flow over a high-lift configuration, Comput. Fluids 39 (2010) 1241–1253.

192. D. König, M. Meinke, W. Schröder, Embedded LES/RANS boundary in zonal simulations, J. Turbulence 11 (7) (2010) 1–25.

193. B. Roidl, M. Meinke, W. Schröder, A zonal RANS/LES method for compressible flows, Comput. Fluids 67 (2012) 1–15.

194. D. Hartmann, A Level-Set Based Method for Compressible Premixed, RWTH Aachen, Dissertation, 2010.

195. D. Hartmann, M. Meinke, W. Schröder, A strictly conervative Cartesian cutcell method for compressible viscous flows on adaptive grids, Comput. Methods Appl. Mech. Eng. 200 (2011) 1038–1052.

196. F. Hausmann, W. Schröder, Coated hot-film sensors for transition detection in cruise flight, J. Aircr. 43 (2) (2006) 456–465.

197. S. Burgmann, Ch. Brücker, W. Schröder, Scanning PIV measurements of a laminar separation bubble, Exp. Fluids 41 (2006) 319–326.

198. S. Burgmann, Investigation of Transitional Separation Bubbles Using ThreeDimensional Measurement Techniques, RWTH Aachen, Dissertation, 2009.

199. Ch. Brücker, J. Spatz, W. Schröder, Wall shear stress imaging using microstructured surfaces with flexile micro-pillars, Exp. Fluids 39 (2) (2005) 464–474.

200. S. Große, Development of the Micro-Pillar Shear-Stress Sensor MPS3 for Turbulent Flows, RWTH Aachen, Dissertation, 2008.

201. S. Große, W. Schröder, Mean wall-shear stress measurements using the micro-pillar shear stress sensors MPS, Meas. Sci. Technol. 19 (2008) 1–12.

202. S. Große, W. Schröder, Wall-shear stress patterns of coherent structures in turbulent duct flow, J. Fluid Mech. 633 (2009) 147–158.

Citations

CHAPTER 1

Dykas, S. and Machalica, D. (2014) The Analysis of the Noise Generation in Gas Turbine Stage. Open Journal of Acoustics, 4, 155-162. doi: 10.4236/oja.2014.44016.

CHAPTER 2

A. G. Sheard and A. Corsini, "The Mechanical Impact of Aerodynamic Stall on Tunnel Ventilation Fans,"International Journal of Rotating Machinery, vol. 2012, Article ID 402763, 12 pages, 2012. doi:10.1155/2012/402763.

CHAPTER 3

Ene Barbu, Valeriu Vilag, Jeni Popescu, Bogdan Gherman, Andreea Petcu, Romulus Petcu, Valentin Silivestru, Tudor Prisecaru, Mihaiella Cretu and Daniel Olaru (2015). The Influence of Inlet Air Cooling and Afterburning on Gas Turbine Cogeneration Groups Performance, Gas Turbines - Materials, Modeling and Performance, Dr. Gurrappa Injeti (Ed.), ISBN: 978-953-51-1743-8, InTech, DOI: 10.5772/59002.

CHAPTER 4

N. Couto, A. Rouboa, E. Monteiro and J. Viera, "Computational Fluid Dynamics Analysis of Greenhouses with Artificial Heat Tube," World Journal of Mechanics, Vol. 2 No. 4, 2012, pp. 181-187. doi:10.4236/wjm.2012.24022.

CHAPTER 5

S. Hady, "A Fundamental Equation of Thermodynamics that Embraces Electrical and Magnetic Potentials," Journal of Electromagnetic Analysis and Applications, Vol. 2 No. 3, 2010, pp. 162-168. doi: 10.4236/jemaa.2010.23023.

CHAPTER 6

Mori, A. and Suzuki, Y. (2013) Grand potential formalism of interfacial thermodynamics for critical nucleus.Natural Science, 5, 631-639. doi: 10.4236/ns.2013.55078.

CHAPTER 7

A. Traverso, L. Magistri, A.F. Massardo, Turbomachinery for the air management and energy recovery in fuel cell gas turbine hybrid systems, Energy, Volume 35, Issue 2, February 2010, Pages 764-777, ISSN 0360-5442, http://dx.doi.org/10.1016/j.energy.2009.09.027.

CHAPTER 8

W. Schröder, Fluid Mechanics Research at the Institute of Aerodynamics, RWTH Aachen University: From 1912 through 2012, European Journal of Mechanics - B/Fluids, Volume 40, July–August 2013, Pages 2-16, ISSN 0997-7546, http://dx.doi.org/10.1016/j.euromechflu.2013.01.005.

Index